How
Plants Get Their
Names

By
L. H. Bailey

Dover Publications, Inc., New York

This new Dover edition, first published in 1963, is an unabridged republication of the work first published by the Macmillan Company in 1933.

International Standard Book Number: 0-486-20796-X

Manufactured in the United States of America
Dover Publications, Inc.
31 East 2nd Street
Mineola, N.Y. 11501

SYNOPSIS OF THE BOOK

which is written for those who may wish
to read it but with the horticulturist
and garden-lover particularly in mind

VI. THE NAMES AND THE WORDS: presentation of lists of names of genera and species with suggestions on pronunciation, the usual meaning or significance of Latin adjectives when employed in botanical binomials, preceded by explanations, all to the end that the reader may find more joy in incorporating names of precision in customary speech and more satisfaction in spelling them.

Begins on page 105

I

ON MY TABLE

ON my table stands a plant richly laden with orange-red cherry-like berries against smooth deep green narrow leaves, a pleasant object on this first day of November when foliage has fallen from the great elms and from the soft maple tree and when signs of approaching winter are on the landscape. It is a stocky arresting plant, good to see. Although only one foot high and about as far across, it holds more than one hundred berries standing boldly upright or sidewise on short stout stalks. The largest oldest berries are about one inch across, nearly globular but slightly flattened endwise; and there are younger berries to make succession. Each berry is held in a fingered cup closely pressed to its rondure.

On the seventh day of last January, in the genial warmth of the greenhouse, a kidney-shaped tomato-like seed was put in the earth; soon a sprout appeared, then aspiring green leaves, and steadily the plantlet grew. Shortly it was pricked off into a two-inch pot, then into a three-inch, then turned from the pot into the open field where it remained in the summer weather; when the cool of autumn came it was lifted from the field and placed in the five-inch pot where it now stands, as I write, erect and handsome.

How and why this plant made itself out of the inert soil and the transparent air I have no way of

1

knowing. It might not be difficult to understand the main physiological processes, but that would not answer the question; we could talk learnedly about heredity and yet not know why it bears orange berries rather than purple, and trough-like leaves rather than flat ones. I do not know why that seed knew how to produce this plant and not a tomato plant,

On my table.—Jerusalem Cherry.

utilizing the same soil and water and atmosphere and sunlight in which, in this same pot, a tomato might have made itself.

Yet I accept the plant for all that. It is mine until its berries fall from age. It is a cheering contrast to the books on the table, the pictures at the side, the calendar that marks the speeding days, and a relief from the ink and paper whereby I write. A mystery is in it that does not pertain to books and tables and deft utensils, an other-worldiness at which we would marvel were it not so familiar. The books stand here year by year with no change except that the bindings loose their grip and fall in shreds by force of gravity, but if I do not water this plant today it will be a wilted wreck tomorrow. I must not let it freeze. For all its apparent ruggedness it is a tender thing, needing the care I give it. Thereby do I have a partnership in it, and a quiet satisfaction.

Despite the regularized treatment it has had, the plant has taken its natural course, with an open horizontally branching informal habit, quite unlike the upright close shapes one sees in the pictures. These formal shapes may be result of clipping or of pinching back, and of other manipulation. There are, indeed, dwarf compact forms that of themselves take a more symmetrical shape, but my plant is not one of them.

It is a woody plant and if I were to grow it as a shrub in the cool greenhouse it would probably become three or four feet high and bear its cherry-berries annually; but the abundance of berries is emphasized on these young pot plants.

On a packet of seeds is the name Jerusalem cherry. It is fitting that the plant should be called "cherry", but why "Jerusalem" should be attached to it I do not know. This part of the name appears not to be very old. I have not come across it in writings of more than about fifty years ago. The plant is not native to Jerusalem and is not mentioned in Post's Flora of Syria, Palestine, and Sinai. Seeds or plants may have been brought from a garden there by someone and the fact recorded in the name. Anything having been picked up in Jerusalem would be likely to bring the name with it.

The name Jerusalem attaches to other plants without special significance. Jerusalem cowslip is a pulmonaria not known in that region except as a cultivated plant. Jerusalem oak is a pigweed with more or less oak-like leaves, native or wild in Africa, Europe and Asia. Jerusalem sage is a phlomis from southern Europe. Two trees are called Jerusalem thorn, one native from southern Europe to China and the other is probably tropical American. Jerusalem corn is a form of sorghum from the Nile region. Jerusalem artichoke is North American, the first name in this case supposed to be a corruption of an Italian word. Of the Jerusalem oak Dr. Prior, authority on names of plants, wrote that "the 'Jerusalem' here seems as in other cases to stand as a vague name for a distant foreign country."

Geographical names frequently go with plants that are native in far distant and different regions. The African marigold of the gardens is from Mexico, as also the Portugal cypress. Cherokee rose, naturalized and widespread in the South, is from China. Arabian jasmine, as well as the Spanish, is native in India. Spanish cedar is native in the West Indies and is not a cedar or a conifer. Peruvian squill is from the Mediterranean region. California pepper-tree and California privet are not Californian. Bethlehem sage is not Judean, Virginian stock is not Virginian, English walnut is not English, Himalaya-berry is not Himalayan, French mulberry is not French nor yet a mulberry.

These various examples testify to the inadequacy of great numbers of English or "common" names of plants. Many of them, as those just cited, are erroneous and misleading. Some of them are duplicates and few

4

of them designate the same plant the world around. Perhaps it is time to start a reformation in vernacular names, or at least to drop many of them from catalogues and books. One cannot "make" common names, although one may coin an English name. A name is not common until it comes into general use. Most plants do not possess true common names.

The plant before us has no good or reasonable accepted common name in English. Florists often cut the matter short and speak of these plants as "cherries". An old English name is "winter cherry", recording the resemblance of the fruits to the cherry and the fact that they persist in winter; it was so known to John Parkinson three hundred years ago; this name is applied also to alkekengi or Chinese lantern-plant as well as to other kinds of physalis, but with less reason. Our plant has also been called "cherry shrub." In other languages the plant has received vernacular names, testifying to its popularity as an ornamental and its long period of cultivation.

One of the interesting old names is "Amomum Plinii", meaning the Amomum of Pliny the Elder who perished in the eruption of Vesuvius in the year 79. Amomum is a Latin (and Greek) name of an aromatic shrub of undetermined identity. Pliny in his Naturalis Historia describes such a shrub, but it is apparently not the winter cherry, as was supposed by the early modern writers. Apparently neither Dioscorides or Theophrastus knew the plant or recorded it.

On a seed-packet is another name than Jerusalem cherry. It is Solanum PseudoCapsicum. This sounds formidable but it has reason and is easily understood.

5

As soon as one finds the word Solanum one knows something about the relationship of the plant, that it has kinship with all other Solanums which include true bittersweet, eggplant and potato, of the nightshade family. The berry of this Solanum is very like the berry or "ball" of the potato.

The item PseudoCapsicum means, of course, false Capsicum; and Capsicum is red-pepper, a closely related plant. The history of the name PseudoCapsicum is long and interesting; although this record is of course somewhat technical, it will reward the reader to follow part of it if he is interested in understanding the delightful old and new art of naming plants.

On my table also is a pepper or capsicum. It is a bonny plant, with brighter livelier colors than the winter cherry. This particular plant has a horticultural history similar to that of the other, and now, in a five-inch pot, it stands nine inches high and about eighteen inches spread, the branches somewhat drooping at the end. The numerous long-conic berries or peppers, about one inch in length, all stand erect above the bright broad leaves, greenish-white at first, then yellowish-white, finally glossy scarlet, the composite making a brilliant contrast; buds and small white flowers are at the end of the twigs. Two good winter window-plants are these, near of kin and closely linked in name.

These New World peppers, very different from those of the eastern tropics from which we obtain the table pepper of commerce, were early introduced to Europe. Peter Martyr writes in 1493 that Columbus

6

had brought a "pepper more pungent than that from Caucasus." In due time these western peppers as a class acquired the name Capsicum, probably from *capsa,* Latin for *box,* because of the box-like soft fruits. To Basil Besler, however, in 1613, in Hortus

On my table also.—Red Pepper.

Eystettensis published probably in Nuremberg, most glorious of horticultural books, they were known as Piper. As botanical names came more and more to be regularized, the word Capsicum was adopted, and it appeared in connection with these plants in the standard work of the great Frenchman, Joseph Pitton de Tournefort, Institutiones Rei Herbariæ, in 1700; from Tournefort the name was taken by Linnæus and is now the accepted nomen of the red-pepper group.

It is pleasant to grow these capsicums, so promptly do they produce their brilliant durable hollow fruits in many shapes and colors. As a garden and field crop they are important, the great puffy kinds for the making of "stuffed peppers" and the smaller more acrid ones for various pickles and seasonings. Of late we have come to grow certain of them in pots for table ornament, and have produced kinds with erect very brilliant peppers that lend lively color to the room; we are indeed clever; yet I find essentially the same kind pictured in Besler more than three centuries ago as Piper Indicum minimum erectum, a name that records the supposed origin of these plants in India. Soon are the histories lost; or, more likely, there was no real history, and in those days geography was not exact. Piper Indicum minimum erectum, the small erect Indian pepper, was apparently prized in the time of Besler, which we think to have been so long ago; and these plants were known to other faithful writers of that period.

There had also come to the gardens of Europe another pepper-like plant but clearly different; this was distinguished from the true Capsicums as Pseudocapsicum or false Capsicum: it is apparently the plant on my table that we call Jerusalem or winter cherry. A very early account of this plant, with picture, was published by the Dutch botanist Rembert Dodoens, Latinized Dodonæus, in his ponderous Stirpium Historiæ Pemptades, published in Antwerp, which in its revised edition of 1616 was quoted by Linnæus. He describes the plant, speaks of its cultivation, explains its name, mentions its medicinal virtues as far as they had been discovered. The full account in Dodonæus is here reproduced, and a free translation of the Latin:

Pseudocapsicum is taller and more shrubby than Capsicum; its stalks are sometimes two cubits long, woody, with numerous branches; leaves oblong, not very broad, smooth, longer and narrower than those of the garden Solanum; flowers white; fruit rounded, red, but paler than that of Capsicum; seed flat, with little or no taste.

An exotic species, cultivated in pots by the Belgians. It is longer lived than Capsicum and can survive for several years if protected from the cold in the winter months.

Pseudocapsicum gets its name from its likeness to Capsicum; there are those who would call it Solanum rubrum or Solanum lignosum but it is not a species of Solanum. The Spaniards call it *Guindas de las Indas*.

Further, it does not agree in temperature with Capsicum; not warming indeed but cooling. What its useful properties are, moreover, has not yet been discovered.

To Basil Besler (if it was indeed Besler who wrote the luxurious Hortus Eystettensis as the title-page avers) a winter cherry was Strichnodendron or nightshade tree; he gave a good description and a picture. The picture is too large to reproduce on this page, for Besler's illustrations are in natural sizes (and thereby do we have a measure of the degree of improvement of garden plants more than three hundred years ago). To Johann Bauhin in 1650 to 1651, the plant was Strychnodendros; and the picture, same size as in his Historiæ Universalis Plantarum published at Yverdun in Switzerland, is shown on page 12. Linnæus recognized the plant as one of his genus Solanum; bringing it into the genus he had the privilege to choose any name for the species itself, but he preferred that of Dodoens: so the plant became Solanum PseudoCapsicum, and this name is now known to all botanists of the world; Linnæus chose to indicate the two elements in the word by writing Capsicum with a capital initial.

De Pseudocapsico.   CAP. XXVII.

Pseudocapsicum.

PSEVDOCAPSICVM altius ac fruti-
cosius est quam Capsicum: caules eius
quandoque bicubitales, lignosi, ramosi: fo-
lia oblonga, latiuscula, læuia, longiora an-
gustioraque quàm hortensis Solani: flores
candidi: fructus rotundus, rubens, dilutiùs
tamen quàm Capsici: semen in hoc pla-
num, nullius aut exigui gustus.

Peregrina etiam stirps, quæ & in fictili-
bus à Belgis alitur. Diuturnioris autem
quàm Capsicum vitæ est, & pluribus annis
superesse potest, si hibernis mensibus à fri-
gore caueatur.

A Capsici similitudine Pseudocapsicum
nomen inuenit: sunt qui Solanum rubrum, aut Solanum lignosum esse velint:
sed Solani non est species. Hispani *Guindas de las Indas* appellant.

Temperie autem Pseudocapsicum cum
Capsico non conuenit: non excalfaciens
siquidém est, sed refrigerans Quæ autem
prætereà eius sint facultates, nondum ex-
ploratum

Pseudocapsicum page of Dodonæus. 1616.

10

That the reader, if he is so inclined, may know how carefully the names and records of plants were built up, even in the time of Linnæus, we may pause to look at his account of *Solanum PseudoCapsicum*. In Species Plantarum ("Species of Plants"), 1753, is the following entry:

PseudoCap- 3. SOLANUM caule inermi fruticofo, foliis lanceola-
ficum. tis repandis, umbellis feffilibus.
Solanum caule inermi fruticofo, foliis ovato- lanceola-
tis integris, floribus folitariis. *Hort. cliff.* 61. *Hort.*
upf. 48. *Roy. lugdb.* 424.
Solanum fruticofum bacciferum. *Bauh. pin.* 61.
Pfeudocapficum. *Dod. pempt.* 718.
Habitat in Madera, ♄

It will be noted that his Solanum no. 3 is described in two lines of Latin, with the specific name Pseudo-Capsicum in the margin. The Latin means that Linnæus had a Solanum with shrubby or woody spineless stem, lanceolate repand or undulate leaves, and flowers in a sessile umbel.

Then follow references to literature. *Hort. cliff.* is Hortus Cliffortianus, a quarto volume by Linnæus, published in 1737, being an account of the plants in the gardens of George Clifford in Holland. *Hort. ups.* is Hortus Upsaliensis, by Linnæus, 1748, an inventory of the plants in his garden at Upsala. *Roy. lugdb.* is Royen, Floræ Leydensis, 1740, an account of plants at Leiden, Holland. In these three works this Solanum was described in similar Latin phrase.

It was named *Solanum fruticosum bacciferum* (shrubby fruit-bearing Solanum) by *Bauh. pin.,* which means Caspar Bauhin's Pinax, published at Basel in 1671. It was Pseudocapsicum of *Dod. pempt.,* which is Dodanæus' Pemptades, as we have already discovered.

11

If we go back to Hortus Cliffortianus we find that Linnæus cites other books, but we need not follow these references except perhaps to mention *Cæsalp. syst.* This reference is to Andrea Cesalpino, Tuscan, whose De Plantis was published in 1583. Cæsalpinius

Strychnodendros or nightshade shrub
of Bauhin, 1650–51.

(as his name is Latinized) was a prophetic man; he was apparently the first to propose a system of classification of plants on the structure of fruits and seeds. He recognized, also, that fossils are organic in origin, and that the heart discharges blood into arteries, in advance of Harvey. In De Plantis, page 215, as cited by Linnæus, is an entry about *Solanum arborescens nuper inter peregrinas allata est,* and then follows a description. This introductory clause or name means a tree-like Solanum that had been recently introduced, among others; Linnæus supposes it to have been the plant he called *Solanum PseudoCapsicum.*

In the Species Plantarum account, that we have reproduced, the last line says that the habitat or place of *Solanum PseudoCapsicum* is in Madera (Madeira Islands); the odd type-character at the end means that the plant is a tree or shrub. The plant grows in Madeira but is said not to be native there; nor do we yet know its nativity. It is ascribed to Brazil, India and other regions. It has been so long in cultivation that it is difficult to say whether occurrences of the plant in fields or open places represent native or run-wild stock. For our purposes, not being here interested in the indigenous habitat, we may well adopt the designation in Index Kewensis (a vast continuing work listing the names of flowering plants of the world), as amphigean, "around the earth."

We have traced the name *Solanum PseudoCapsicum* but on another seed-packet before me are the names Jerusalem cherry and *Solanum Capsicastrum.* If we were to plant the Capsicastrum seeds we should undoubtedly obtain plants like those from the Pseudo-Capsicum packet. Here we are thrown completely off the track and we may not get back on it again by the discussion on page 63.

Now may we return to our capsicum, still standing on the table, unmindful of all this ink. We left it as Capsicum Indicum of Besler. Linnæus accepted the genus Capsicum in Species Plantarum, and described two species, *Capsicum annuum* (annual capsicum) and *C. frutescens* (shrubby capsicum). The former he ascribes to tropical America and the latter to India. The two species are published together in the one account, but *annuum* stands first, and in case of doubt this name, under the rules, is to have precedence over the other. It has been the custom to call the peppers of northern gardens *C. annuum,* assuming them to be distinct from the shrubby or woody kinds. The shrubby kinds look distinct enough when one sees them in the wild in hot countries; once I cut a durable cane from the hard dense wood of a pepper bush that was higher than my head; yet the herbaceous and the ligneous kinds, I am convinced, are all one thing. Much experience in growing them confirms me. To this effect I wrote some years ago, and the paragraph explains another change of names of the kind that bothers the plant-grower:

"I am convinced that the horticultural kinds are all forms of one species, and that the species is shrubby, the herbaceous or so-called annual kinds being races that develop in a short season and do not become woody before killed by frost. In the Capsicum shrubs of the tropics one finds puffy fruits of the bell-pepper type as well as the slender finger-like and the berry-like kinds; and when the northern kinds are grown in the tropics they become shrubs. Leaf variation also has equal range. I therefore propose to arrange the most significant forms of this multifarious species under *C. frutescens* rather than under *C. annuum.* In doing so, I accept the second rather than the first of the two names proposed by Linnæus in Species Plantarum; but when no question of authority or priority is involved, I cannot allow the accident of precedence on pages to obscure a biological fact."

On my table the two plants stand, one at my left hand, one at my right. Beyond the window-pane the chill of late autumn frost is in the air. Proud herbs of summer are collapsed. Brilliant lilac colchicums are gone. The crimson habranthus by the door has passed another year. Wilted petunias still hold a waning bloom. A clump of autumn bugbane tries to defy the frost, and the faint yellow of little-flowered chrysanthemums I found many years ago on far hills of China yet shines in the border. Insects are covered or gone. Birds of summer have flown. Sparrows will chipper at the eaves; the small flock of starlings will gather in the top of the great hickory tree. Soon the twigs of bushes will be laden with snow. Yet here my potted plants are lively and brilliant with the sunshine of milder climes. Centuries ago the seeds were brought by somebody from somewhere, and in all the eventful generations the plants hold true to their type; one is still a solanum and one is still a capsicum; they carry the peculiar features that were developed in untold cycles of time.

The plants represent the round world to me. They are reminders also of careful observers hundreds of years ago who left good records in aristocratic Latin when the common vernacular language was considered not to be sufficient medium for such learning. Centuries are tied together.

II

LINNÆUS

IT is profitless to go farther in quest of names until we know Linnæus.

Carl Linnæus was born in southern Sweden in 1707. His father, Nils Ingemarsson, took a Latin surname when he began his school and university career to become a scholar and eventually a churchman, adapting it from a certain famous lind, the lime-tree or linden. It was custom in those days for persons to choose a Latin name or to Latinize the patronymic. The family of the cousins of Linnæus chose the name Tiliander from the same tree, Tilia being Latin for the lindens. Another branch of the family became Lindelius. The particular lind tree, it is written, "had acquired a sanctity amongst the neighbours, who firmly believed that ill-fortune surely befell those who took even a twig from the grand and stately tree." Even the fallen twigs were dangerous to remove, and they were heaped about the base of the tree. It had perished by 1823.

To the people the name Linnæus was rendered Linné, the accent preserving the essential pronunciation of the word. Linnæus wrote that "Linnæus or Linné are the same to me; one is Latin, the other Swedish." His great Latin books were written naturally under the name Linnæus, and thus is he mostly known to naturalists. In later life a patent of nobility was

LINNÆUS.

granted him and he was then Carl von Linné. We find him signing himself as Carolus Linnæus Smolander, his province or "nation" being Smoland and Carolus being the Latin form of Carl or Charles; also as Carl Linnæus, Carl Linné, and Carl v. Linné. This much is by way of preface to explain the forms in which the name of this marvellous man appear.

Our interest in Linnæus is in relation to natural history. This relationship cannot be fully appreciated without knowing something of the state of natural science in his day, and of the social expression of the people. It would be too much to undertake such an inquiry; but it may be said that there was scarcely an independent science of botany in that epoch pursued for its own sake but only as a department of medicine; and public opinion was not free to allow the pursuit of knowledge in any or every direction.

The young Linnæus, therefore, made his way with difficulty and with few established aids. Yet he became outstanding authority in what were called the three kingdoms of nature—plants, animals, minerals. He was an extensive field naturalist, and skilled also as an assayer. His chosen profession in early life was medicine, not considering himself qualified for the church; but he became professor in the university at Upsala and there attracted great numbers of students from many parts of the world and trained naturalists who traveled to far parts for natural history specimens, as Thunberg to the Cape of Good Hope and Japan, Kalm to North America, Loefling to Spain and South America, Forskal to Egypt and Arabia. As his knowledge was comprehensive so was his enthusiasm unbounded, and the influence on students was commanding.

Linnæus was interested primarily in botany, not only in the kinds of plants but in their distribution and natural history. The knowledge of plants had been accumulating for some centuries and it was preserved in many tomes largely, of course, in Latin which was the language of learning. Yet the knowledge lacked system because there was no adequate plan of arrangement and no simple set of names. It was in these two fields that Linnæus made his outstanding contribution to biological science—in classification and in nomenclature.

Before his time the classificational schemes of Ray and Tournefort were in vogue. The work of Tournefort was nearly contemporaneous with Linnæus, his great Institutiones Rei Herbariæ, in three volumes, having appeared in 1700; he died the year after Linnæus was born. The title of this important Tournefortian work is hardly translatable into current English; perhaps it will suffice, for descriptive purposes, to call it "Principles of Botany" (herbaria: knowledge of plants, or botany). Two of the volumes are devoted to engravings of plants that in precision and beauty would do credit to books of our own time. Tournefort knew about 10,000 plants. These could be divided into trees (or big woody plants) and herbs, and these divisions separated into groups that bore petals and those that did not (petaliferous and apetalous) and further on the shape of the corolla. He had no names for plants in the modern sense but called them mostly by Latin phrases or clauses, as we shall presently see. He established the concept of the genus, that in maturer form has come down to the present day.

Sexuality in plants was not accepted by Tournefort, although the idea had been cogently advanced. It was taken up by Linnæus, however, who, in his examination of stamens and pistils in verification of the proposition, hit upon the plan of using them as a basis of classification. This great Linnæan system, although destined to be overthrown as the author of it himself had foreseen, enabled the plants of the world to be ranged in definite classes and orders, and it at once brought confusion into symmetry. There are thirteen classes based on the number of stamens, from 1 to 11, then 20, then many; two on relative lengths of stamens; four with connected stamens; one in which stamens and pistils (or styles) are consolidated; three with imperfect flowers, as monœcious, diœcious, polygamous; one without these organs, the cryptogams; this makes twenty-four classes. The classes are further divided into orders on the number of pistils or of styles.

This "sexual system," as it has been inappropriately called, actually brought together great numbers of plants closely related, and, on the other hand, it also divorced many natural relationships. By bringing order from scattered records it made the kinds of plants more available for study and comparison, and prepared the way for the natural families perfected by the Frenchmen Jussieu, uncle and nephew, and by Adanson, and by others to our own day.

The literary labors of Linnæus were phenomenal. He wrote about one hundred and eighty books, some of which were published after his death which oc-

curred in 1778. The books that interest us particularly
in the present narrative are Genera Plantarum
("Genera of Plants") which appeared in 1737 and
went through several editions, and Species Plantarum
("Species of Plants") in 1753. "Some 7300 species
are diagnosed in this work," according to Ellison
Hawks in his Pioneers of Plant Study, "with their
synonymy and localities—arranged, of course, according to the Sexual System. Although the number
is less than those described by Tournefort or Ray, almost all had been examined by Linnæus himself and
were represented in his herbarium." The work
runs to 1200 pages, aside from indexes, bound in
two compact octavo volumes, published in Stock-
holm. It is now rare, but a facsimile reproduction
by photo-engraving is available. Subsequent editions of the Species by himself appeared in 1762–3
and 1764.

In the Genera, the concept of the genus was defined and recorded essentially as we know it at the
present day; in the fifth edition, 1754, which is the
most important issue for purposes of nomenclature,
1105 genera are described. In the second work, Species Plantarum, all the species of plants known to
him at that time were described under the appropriate
genera; and in the margin he gave a specific or index-
ing name, as we have observed in the case of *Solanum
PseudoCapsicum*. He also developed the concept of
varieties subordinate to species and entered varietal
names in the margin in a different type. Subsequently
he did similar service for animals.

Genera, species, varieties, these are the three categories of the forms of life, definitely stabilized by
Linnæus, and these denominations we must under-

stand before we can undertake the study of the kinds of plants and animals or approach the subject of nomenclature.

Pyrus is the genus of the pome fruits.

Malus is the apple species.

paradisiaca is a variety of the apple.

In writing this becomes

Pyrus Malus, the apple.

Pyrus Malus var. *paradisiaca,* the paradise apple.

If we omit the species-name (or specific name), Malus, and write *Pyrus paradisiaca* we commit two errors: we make a new name, and we assert that the paradise apple is not a variety of the apple species but a separate species by itself, of distinct genesis in nature. It is exceedingly important that we do not confuse the concepts of species and variety, else we cannot speak and write of plants with discrimination.

It is impossible accurately to define what is meant by species. The naturalist gradually acquires the idea and it becomes an unconscious part of his attitude toward living things. Nature is not laid out in formal lines. Perhaps it will aid the inquirer if I repeat the brief definition I wrote in Hortus: A kind of plant or animal that is distinct from other kinds in marked or essential features, that has good characters of identification, and that may be assumed to represent in nature a continuing succession of individuals from generation to generation.

Even as simple a statement as this cannot be understood merely by reading it. The meaning gradually comes to one. The apple is one species, pear another, belonging to what Linnæus considered to be a single genus or group; they are not varieties. If the reader wishes to go farther in this subject he may look up in Hortus the entries family, genus, variety. Let it be said, before we leave the subject, that the word species is either singular or plural: we speak of one species or of six species. When only one is meant, I have seen it written specie: but that is quite another affair, representing certain interesting pieces of metal I have known other persons to have in their pockets.

To the pre-Linnæans plants had no accepted or uniform short definite technical names. Thus to Gronovius (1739–43), Royen (1740) and others catnip was "Nepeta floribus interrupte spicatis pedunculatis," which is a brief description of the plant; Linnæus described it under Nepeta and put *cataria* in the margin, making the name *Nepeta cataria,* as we have it now (*cataria,* a late Latin word, "pertaining to cats").

To Johann Bauhin the watermelon was entered as "Citrullus folio colocynthidis secto, semine nigro"; Linnæus placed it in his genus Cucurbita with *Citrullus* in the margin, and the plant to him was *Cucurbita Citrullus.* The carnation in several works was written "Dianthus floribus solitariis, squamis calycinis subovatis brevissimis, corollis crenatis," which is a beautiful characterization; Linnæus made it *Dianthus Caryophyllus.*

These descriptive phrases as designations of plants seem strange enough to us, and bungling. But not all of them were so long. Before me is a pre-Linnæan designation of a maple: *Acer orientalis, hederæ folio,* oriental ivy-leaved maple. Then I pick up a current American nursery catalogue and find: *Acer polymorphum dissectum pendulum.* I have grown a flower-garden poppy under the name *Papaver*

Rhœas coccineum aureum, and a phlox as *Phlox Drummondii rosea alba oculata.* Perhaps the botanists of a few centuries ago did not have so much trouble with the titles as we imagine, particularly as they knew Latin.

It came that plants acquired two names, one representing the genus or family group, as Johnson is a family name, and the other the particular species. This is *binomial nomenclature,* by means of which all plants and all animals are known by all people in all countries who speak or write of them with precision. As a system it begins with Linnæus in Species Plantarum in 1753: that date is the starting-point for the naming of plants; the starting-point for animals is 1758, in the tenth edition of Linnæus' Systema Naturæ ("System of Nature"). In fact, however, Linnæus had employed specific names as early as 1745 in the index of a Swedish book recording his travels in the provinces Oland and Gothland, but they had not then become a system; and there are descriptive paragraphs in his Hortus Cliffortianus, 1737, headed with a binomial: Capsicum annum and Capsicum frutescens are examples. Again, a single word was used as a specific or trivial name in volume two of his Amœnitates Academicæ, 1749.

We must not conclude from the foregoing discussion that two-word designation of plants was unknown before Linnæus. Open on my table is a choice vellum book of the Frenchman Carolus Clusius (whose name in French was L'Ecluse or L'Escluse or Lescluze), printed in 1576, on his botanical observations in Spain; here is a picture named Genista tinctoria, another titled Dorycnium Hispanicum, and many others. These names were not part of an organized system, however; many of the plants were known by numbers, as Cytisus I, Cytisus II, Cytisus III, Cytisus IIII. These cases, and others that might be cited, show that nomenclature began to take form early in the modern historical period.

Then, as now, were there earnest enthusiastic students of plants, whose devotion would do credit to the best intentions of this our luxurious day. Read the paragraph about Clusius by Benjamin Daydon Jackson, recent master historian of botany: He "was almost as much distinguished by his personal misfortunes as by his sterling botanical merit. He travelled through Spain to observe the plants of that peninsula, and Hungary and Bohemia for alpine plants; in doing so he suffered greatly from accidents which one after another happened to him, and at length quite crippled him, but failed to quench his unappeasable ardour in the pursuit of the knowledge of plants. His Latin style is much praised for its purity, and as he was first to describe a very large number of new plants, his books are of great interest. He ended his days as professor of botany, at Leyden, in 1609."

A binomial is not only a name of a plant: it also places the plant in a system, and adds associated interests. Thus, when Linnæus named the winter cherry he related it to the potato, tomato, and the nightshades by placing it in the genus Solanum; he also associated with it the old Pseudocapsicum history: so that *Solanum PseudoCapsicum* is much more than a nom. This is true of all binomials by whomsoever made. When Michaux in 1803 "made" the species *Rhododendron catawbiense* he classified it by the act of putting it in the genus Rhododendron and also recorded the Catawba region where he collected it,—"in montibus excelsis Carolinæ septentrionalis juxta originem amnis *Catawba*," in the high mountains of North Carolina near the head waters of the Catawba River.

The generic name is always part of the binomial: *PseudoCapsicum* is not sufficient to designate the winter cherry nor does *catawbiense* alone identify the rhododendron. If the plant is subsequently placed in another genus (for reasons yet to be disclosed), then the acquired genus lends its name: thus Pursh in 1814 described the species *Azalea arborescens,* the arborescent or tree azalea; but Torrey thought the azaleas should not be botanically separated from the rhododendrons and in 1824 he made the binomial *Rhododendron arborescens;* if this disposition is accepted, Pursh's name becomes a synonym. In 1894 I founded the species *Prunus Besseyi,* the western sand cherry, until that time not recognized as distinct from other native cherries, naming it in compliment to my illustrious friend and mentor, the late Charles E. Bessey; in 1898 Smyth put it over into the genus Cerasus as *C. Besseyi,* thinking the cherries to be so distinct from the plums as to merit a genus of their own, Cerasus having good history as a generic name: if one places all these stone-fruits in Prunus, *Cerasus Besseyi* becomes a synonym; if one prefers to adopt Cerasus, then *Prunus Besseyi* becomes a synonym.

As the species is subordinate to the genus, so is the variety subordinate to the species. *Fraxinus excelsior* is the European ash; *F. excelsior* var. *asplenifolia* is a form or kind of *excelsior;* sometimes such names are written without the abbreviation var., and we have then a straight trinomial, as *Fraxinus excelsior asplenifolia,* but the sense or significance is not altered thereby.

The system of binomial nomenclature is one of the best inventions of men. It is effective; it is beautiful in its simplicity. It serves all men and women. It is endlessly extensible. It answered the purpose of Linnæus and his associates when the number of known plants was few; it is in daily use one hundred and eighty years later, when plants are numbered in the hundreds of thousands. It is similarly in use in the animal kingdom; the system served for the 4,236 animals named and described by Linnæus; it applies to-day for all the animals known to men, including the hundreds of thousands of insects.

Every binomial has meaning; it is significant. To know the names of the forms of life is one of the keenest of satisfactions; it brings one into relationship with living things, in endless variety; it multiplies the contacts.

We have seen how Linnæus harvested the extensive records of his predecessors. Most of these antecessors are known to us as herbalists, persons who wrote of plants primarily in respect to their virtues in the art of healing. But some of them, as Tournefort, were interested directly in the study of plants with a view to identification and characteristics, much as the modern scientific spirit impels. Thus, in his account of Geranium in Classis VI, which includes herbs and subshrubs with rosaceous flowers, the genus is described in six Latin lines, and then follow eighty-one "species" (rather, kinds) as described but not named in preceding literature, all without reference to "vertues."

When Linnæus established his genus Geranium he cited Tournefort's plate, and then proceeded with a regular written diagnosis; he accepted thirty-nine species, some of which are now placed in Erodium and others in Pelargonium. Linnæus did not always accept the generic names of Tournefort.

The reader may wish to see some of the plates in Tournefort. We may begin with Corona Solis. It will be recognized that here we have the sunflower; Linnæus did not adopt Corona Solis although he cites the plate; he makes the genus Helianthus, Latinized from the Greek *helios,* sun, and *anthos,* flower; and thus do we say it to the present hour.

Tournefort's explanation of his plate of Corona Solis will interest us. At A is the radiating flower, the disc indicated by B; one of the many florets or flosculi is at D, with the embryo (fruit) at E; a neutral floret is at G, with its great corolla or ray F; at I is the calyx (involucre), and below are details of floral parts. At C is the true corona, the crown of Sol the sun.

Again, we may choose Avena, the oat, a beautiful picture; Linnæus accepted the name from Tournefort. At A are shown the many flowers in the "calyx" D; BC is a stamen, E pistil, and G the "seeds"; at I are fascicles, and they are combined in the long spike marked midway by HH. Once again, we may look at Lycopersicon the tomato, well shown in detail of fruit and flower; note that the flower, even in that early day, carried more than the normal five corolla-parts and calyx-parts, seen entire in CA, with corolla removed in CD, back view at AB, front view at A. The whole fruit is at E, in section at F (and the many cells may be noted), seed at G.

Corona Solis

Tournefort's sunflower. 1700.

29

Linnæus placed the tomato in Solanum, along with Tournefort's Melongena or eggplant, one becoming *Solanum Lycopersicum* and the other *S. Melongena.* Philip Miller, contemporary of Linnæus, kept the tomato and a few related species in a separate genus and under his treatment the plant became *Lycopersicon* or *Lycopersicum esculentum;* and this is the binomial under which it is now known although one of the current authorities re-unites it with Solanum. The lobulate tomato fruit in Tournefort is now seldom seen in the United States, the larger or more uniform "smooth" fruit being preferred; but this flat creased tomato was frequent when I began work on tomatoes now well-nigh fifty years ago; I still see it commonly in the tropics. It was not until Waring introduced the Trophy in 1870 that the modern race of North American tomatoes began rapidly to displace all others, with the development of commercial vegetable-gardening. I remember the interest it aroused.

Thus, now, have we made brief acquaintance of Linnæus, sometimes known to moderns as the "father of botany" because plants cannot be conveniently studied and records made of them, whether in anatomy or physiology or genetics or taxonomy, until we can call them by name. His was a systematic synthetic mind. He united the scattered essentially unclassified records of centuries. He brought order into the study of plants. This order was particularly needed at that epoch when the expansion of trade had begun to bring strange and numerous plants from many parts of the world.

From this small account it is evident that Linnæus

Avena *Aveine*

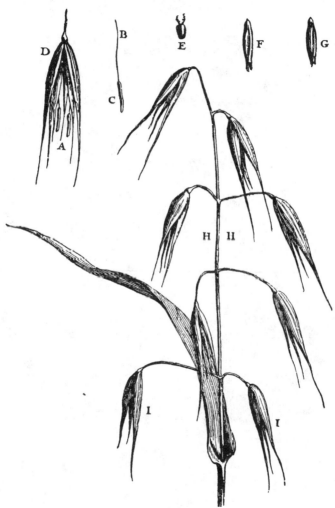

Tournefort's oat. 1700. Latin, *Avena;* French, Aveine (Avoine).

Lycoperſicon

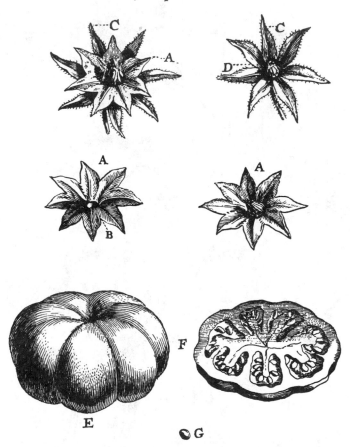

Tournefort's Lycopersicon or tomato. 1700.

had a passion for arrangement or system. He systematized everything, a necessary process to bring together the accumulated records of centuries and to place them in orderliness. He was a synthesist, as Darwin in quite another field was a synthesist.

Linnæus was a systematist in natural history. When we speak of a systematist in zoölogy or botany we designate one who studies the kinds of animals and plants, naming and classifying them. In plants this field is called systematic botany, a cumbersome dubious term that should fall into disuse. Sometimes the subject is known as taxonomy, but this term signifies classification only. Perhaps we would do well to speak of this science as systematics, as we have mathematics, and the devotee of the subject is a systematist.

Systematics is oldest of the botanical sciences, and also, as we shall see, still new and commanding. The subject is as fresh and compelling as when Linnæus tramped the fells of Lapland or strode the fields of Sweden.

III

IDENTIFICATION

FROM a catalogue I ordered seeds of *Cleome gigantea,* spider-flower. The seeds produced *Polansia trachysperma,* clammy-weed.

Both these binomials are correct; they have regular botanical standing. They are accepted nomenclature; but identification of the plant was erroneous.

Seeds were purchased at cassabanana, that bears an ornamental gourd-like fruit; binomial name of cassabanana is *Sicana odorifera.* The seeds yielded wax gourd, *Benincasa cerifera.* Again both names are correct and the plants were correct, but the seed-packet was in error. Systems of nomenclature do not correct seed-packets.

Two plants are known as babys-breath. One is *Gypsophila paniculata,* of the pink family. The other is known in horticultural literature as *Galium Mollugo,* of the madder family. Both are in common cultivation. But it now transpires that the galium has been misidentified and the plant in gardens as *G. Mollugo* is really *Galium aristatum;* but that is not the fault of nomenclature.

The columnar Greek juniper is *Juniperus excelsa* var. *stricta;* but the plant sold under this botanical name in the North is *Juniperus chinensis* var. *pyramidalis.* Both names are correct by all the systems and rules.

34

A binomial or trinomial is of consequence only when applied to the plant to which it belongs and to none other. That is, nomenclature follows identification.

The first problem, then, in clarifying the names of cultivated plants is to identify the plants to be named. This fact is not sufficiently appreciated by plantsmen. We may make endless rules and standardized lists and yet names of plants may lie in confusion because the plants are confused.

Identification is a primary necessity to the understanding of the world. We must accurately identify heavenly bodies before we can chart and study them. We must identify clouds if we are to understand the atmosphere and meteorology. The engineer identifies every element in a machine. The geologist knows his rocks by name. Chemists know and name the substances and reagents with which they work. The entomologist actually knows his insects before he attempts to combat them with much hope of success. The zoölogist knows his animals without guessing, and the botanist his plants; then only can he give them names. The historian identifies his events and the records of them. The physician is skilled in identifying symptoms. Any competent person is able to identify emotions and perhaps to classify them. To identify is a fundamental educational process.

So, then, if the plant-lover wishes to have accurate stabilized names for his plants he must be sure that his plants are the ones to which the names apply. He acquires this knowledge by experience; but it is a sad fact that error is acquired as rapidly as verity; perhaps it is acquired more readily because it does not demand proofs. A person may grow a plant for years under a given name and yet he may have the wrong plant. Gardeners rely on the label by which the plant is received; yet the label may not be reliable.

All this is not to suggest that plants commonly are erroneously determined; yet error in this respect is common enough in horticulture to present a real problem. The first requisite on the part of the grower is to know plants critically, to see differences and the minor marks of identification and to be able also to test his observations against technical descriptions in reliable books. This means a desire to know plants thus intimately; this is an essential preparation for real gardening; it yields one of the best of satisfactions, when one is able to see.

It is by no means always the nurseryman's or the seedsman's fault that his plants or seeds are misnamed. He, in turn, accepts the stock as he receives it from reliable sources. Some kinds of plants are very difficult to distinguish from related kinds. In many cases botanists themselves are not certain. There are cases in which plants have been in cultivation for generations under erroneous names, and have been so accepted in the best books. Thus, for example, the common little narrow-leaved flowering-almond of gardens and yards, in many double forms, was long known as *Prunus japonica,* but it now transpires that it is mostly *Prunus glandulosa,* the true *P. japonica* being less frequent. It is worth pausing a moment to see how this case works out.

In 1784 Thunberg the Swede, successor to Linnæus

and who, as we have noted, had travelled in Japan, described two dwarf prunuses, *Prunus japonica* and *P. glandulosa*. Subsequent authors supposed them to be the same, and the stock in cultivation came to be called *P. japonica*. When Emil Koehne took up the study of the prunus specimens collected in the Orient by the late E. H. Wilson he wrote in 1912 of *Prunus glandulosa:* "For a century this species has been always confused with *P. japonica* Thunberg, but it is very distinct and not connected with the latter by any intermediate forms." He therefore pointed out the differences between the two species. In the Cyclopedia of American Horticulture, 1901, the plant is entered as *P. japonica;* in its successor, the Standard Cyclopedia of Horticulture, 1916, both species are entered and contrasted; subsequent observation indicates that *P. japonica* is apparently not as common in cultivation, at least not in the East, unless in test-grounds and botanical collections. If the gardener is distressed because names have been changed he should also be comforted by the fact that we have learned something: we have two of this type of dwarf flowering-almonds rather than one.

If the reader is not in too great haste to be up and away we shall pause still another paragraph on this interesting prunus case. It is simple enough for a plant-grower to call any prunus coming from Japan *Prunus japonica;* thus it unfortunately happens that the name has been applied in horticulture to the Japanese plum, for which the correct binomial is perhaps *Prunus salicina;* also to the pendent form of the rosebud cherry, *P. subhirtella* var. *pendula:* that is to say, *P. japonica,* Hort. is a synonym of both these names, but *P. japonica,* Thunb., is a good species by itself.

Also the name *P. glandulosa* is confused: Torrey and Gray applied this name in 1838 to the little "wild peach" of Texas, probably unaware of Thunberg's nomen. In 1840 Hooker placed this Texan plant in Amygdalus, a genus we shall meet again before we leave this book in connection with the peach. In Amygdalus the name *glandulosa* may stand, there bring no earlier *glandulosa* in this genus; but in Prunus the name cannot hold for the Texan plant because of the earlier *glandulosa* of Thunberg; to avoid the duplication in Prunus, Camillo Schneider in 1906 proposed the binomial *P. Hookeri* for the Texan plant, but it turns out that as early as 1843 Dietrich had made the name *Prunus texana* for the species, and by priority this binomial must hold if the bush is retained in the genus Prunus. Yet again: to Asa Gray in his long-popular Field, Forest and Garden Botany, 1868, our little flowering-almond was known as *Prunus nana;* when I revised that book in 1895, I was able to say that the true *P. nana* is quite another plant, and entered the narrow-leaved flowering-almond as *P. japonica,* "generally, but erroneously, called *P. nana* in gardens." We may add, also, that in gardening literature the name *Prunus sinensis* has been unauthoritatively applied to the species-group *glandulosa.* Let us hope that we finally have it correct: from 1784 to 1912 is not a long epoch for error to be in the process of solution, seeing that the world is yet ever so young.

In some cases a species started in confusion, without clear concept of a unit or type. *Iris germanica* is an example. It is a mixture or at least indefinite, probably even in the time of Linnæus a series of garden forms. There is no specimen in the herbarium of Lin-

38

næus bearing his identification, although there is one by his son. It is unknown in a native state. What to do in a case like this is to do the best we can. In some cases the nom is disregarded, as a *nomen incertum* or *nomen dubium* (uncertain or doubtful name). Sometimes it may be accepted for a certain plant by common consent, even without typification, but this practice is allowable, if at all, only in historic cases.

When errors are discovered and corrected as the result of identification, the horticulturist is not to complain that names have been changed: the plant has finally been properly determined, and he should be thankful. The accumulation of knowledge is a process of eliminating errors. We hope the process will not fall into disuse.

The naming of plants under rules of nomenclature is an effort to tell the truth. Its purpose is not to serve the convenience of those who sell plants or write labels or edit books; it is not commercial. Serving the truth it thereby serves everybody. In the end, nomenclature rests on the plants rather than on printed regulations.

In many or even in most cases the gardener himself cannot make sure of the identity of doubtful plants. He refers the case to one who knows. Unfortunately, there are none too many persons who are critical students in this field, and there seems to be no general desire in the United States for accurate determination of horticultural plants. This desire is active in wild or native plants.

There are two great aids to the determination of

plants, the botanic garden and the herbarium. Botanic gardens may abound in horticultural plants and herbaria usually lack them; yet the competent herbarium is indispensable so far as identification is concerned.

Plants subject to removal, to death and the substitution of others in their places, to carelessness of workmen with labels, to interference by visitors, to loss of numbers and tags, may readily become mislabelled. Botanic gardens exercise great care to keep plants properly labelled, but shifts and accidents occur in spite of oversight. Moreover, not nearly all the kinds of plants can be grown in any one botanic garden or be in condition for study at the same moment. Limits are set by acreage, cost, soils and climate. Of course the botanic garden has other great merits aside from accurate naming, if it is a scientific institution, but with these services we are not for the moment concerned.

An herbarium is a collection of dried plants. The plants are dead, perhaps for a hundred years; therefore the horticulturist may hold them in high contempt. Persons always ask whether such subjects keep their color; perhaps not; they are not made for looks in the gardener's sense: they are records. Yet they have a fitness and beauty all their own if properly prepared, preserved and housed; and anything not thus conscientiously wrought is likely to have slender value and certainly no attractiveness. Herbarium specimens are not souvenirs.

When an herbarium specimen is once properly placed on adequate paper and determined as to species or variety, it constitutes a practically unchanging record or evidence by means of which other plants, living or dead, may be compared and verified. For be it known that the essential marks of difference between plants are retained in these cabinet specimens.

The specimens are "mounted," in the large herbaria, on sheets of strong white paper by being glued fast; the paper size in North America is 11½ × 16½ inches, called "sheets." These sheets are placed mostly several together in strong heavy folders known as "covers." The covers are filed flat in inclosed pigeonholes. Nuts, cones, and the like are kept in boxes or other containers, and big soft fruits in liquid or represented by photographs.

The reader is already asking how long herbarium specimens will keep. We cannot yet answer that question because they have been made only a few centuries. The herbarium of Cæsalpinius, who died in 1603, is preserved in Florence. The question is, how long the paper will last. Bugs like these specimens and spend all their lives in them, becoming pulpy and juicy on materials that have been as dry as a manuscript for no end of time. If bugs are kept away, and damp and dust, and other proper care extended, these records are as permanent as most others that men make laboriously. Recently I received mounted specimens made by John Stuart, third Earl of Bute, Prime Minister, who died in 1792; plants and paper are attractively preserved.

The herbarium is for identification and record. If there is a growing collection in connection with it, much will be gained; and a library is essential. It is at such places or institutions that the horticulturist as well as the botanist may expect accurate determinations to be made.

How to send material for identification requires a few paragraphs. At the start it is to be understood that many species of plants are so much alike that ample specimens are required to expose the differences. These dissimilarities may be in foliage, flower-bearing habit, flowers, pods, seeds; often the underground parts are characteristic. The larger the piece sent to a botanist, within decent limits, the easier it is for him to make determination and the more certain will be his findings. It is not fair to ask a person to spend time on fragments and unrelated pieces.

Specimens should be flat. Do not roll them; by the time folded fresh material reaches its destination it is likely to be in pieces or so curled as to be impossible of straightening out. Do not wrap in cotton or in excelsior or in moss; it is not right to impose on the recipient to pick out the stuff and to get the specimens untangled and straight.

The best material is that which is pressed flat, so that it may go on an herbarium sheet if necessary. The size need be no larger than an herbarium sheet (roughly 12 × 17 inches). It may be sent green, unless the distance is very great, between good thicknesses of soft paper (as newspaper), with stiff flat cardboard top and bottom, *tied tight, kept flat,* wrapped securely and sent by mail. Be sure that the living specimens are dry when put in the papers.

If the distance is so great or the material so soft or fragile that it is likely to mold in transit, specimens should be regularly pressed and dried, with frequent changes, before shipment is made.

Herbarium sheet of Rosa, in flower and fruit. Nearly one-third full size.

There may be exceptions to this procedure. Stiff things, like pine and spruce branches, may be sent in boxes, with the cones. Big fruits and nuts are also mailed in boxes. If it is desired to show blossoms in full natural condition, the material may be dispatched as cut flowers or pot plants are handled, but this is seldom necessary.

Put labels or tags with the specimens. The recipient will be aided by any information about the plants, as dates, stature, whether wild or cultivated, and if native then the habitat.

In other words, take pains in procuring and sending the material.

New species of plants are founded on dried specimens. When a botanist returns from collecting in some far place persons ask him at once whether he found any new (undescribed) species. He does not know. He must unpack his dried specimens and assort them; these specimens must be studied by special students of the groups, orchids being sent to one person, ferns to another, sedges to another, grasses to still another. Comparisons must be made with all other similar plants already preserved somewhere; literature must be consulted. Weeks or months or years afterwards the collector may be able to answer whether he has novelties.

If a species new to science is fortunately found among them, the description is drawn from the dried material, and the particular specimen is preserved as a "type," available for any competent person to examine in any of the years to come. No man or woman now "makes" a new species without preserving a type specimen as evidence.

In some cases, however, the process is reversed.

The person may be a student of a particular group of plants, knowing them all. He visits a new locality, and practically at sight recognizes undescribed species of the group or genus. But just the same he makes specimens and securely preserves them, all the more religiously because they are his favorites and he carries special responsibility.

Often new species are discovered in the herbarium itself. In these days one specimen of a species is not sufficient. The genus must be represented by material from many different regions to show range and geographic distribution and to exhibit variation. If a botanist says that a certain singular plant is native in Michigan or Alabama, it is expected that he has a specimen to prove it. When many specimens are assembled of a supposed single species it may be found that very distinct plants are involved, with consistent ranges. My first species, published in 1884 along with others, was a sedge discovered in mounted herbarium specimens; the plants were so much alike that two species were mounted on a single sheet; I separated one as *Carex multicaulis,* and it was years later that I first saw it in nature from the saddle on slopes of Mt. Shasta. Strangely enough, my latest species is also a segregate from scores of herbarium sheets (although I know it also in the field) "made" this very day and named *Rubus abactus,* not published as this is written, native in many places, as the specimens disclose, in the northeastern United States.

Confronted with a new species, the botanist or zoologist has choice of any name not before applied in that genus. It should, of course, agree with its genus in verbal Latin form; he may choose to commemorate a person who aided him, a fellow collector, or record the place or habitat; he may prefer an adjective descriptive of some feature or "character" of the plant. Once made and published, the name cannot be changed by himself or anybody else although it may not be adopted by others.

The herbaria of the world are the records of the plants so far as known. They are huge card indexes, with the plants glued on the cards. They are the conservators of the knowledge of the vegetation of the earth. Every year their value naturally increases.

The value and also the interest in these herbarium sheets lies in no small degree in the labels that accompany the plants. They make note of many lands; they are reminders of many collectors, dates perhaps long ago, lands on which the plants grew, all at one's command equally in the height of summer or the deeps of winter, in days of driving storm when one may travel indoors. In no way, perhaps, in such small compass does one condense so much human interest.

Before me is a cover of *Thymus Serpyllum,* an aromatic ground-cover known to gardeners as mother-of-thyme. We may pause to enjoy the name Serpyllum: related to a Greek word signifying *creep* or *creeping,* allied to *serpent,* taken into Latin as a name for the wild thyme, employed by pre-Linnæans as a generic name or substantive for a group of plants, in English still preserved as serpolet which is a name for creeping thyme, and given us permanently by Linnæus as the specific name of this particular Thymus; suggested in later time by names of small-leaved or thyme-leaved plants in other genera, as in the weedy sandwort *Arenaria serpyllifolia,* one of the bluets

Houstonia serpyllifolia, well-known synonym for the artillery-plant of greenhouses *Pilea serpyllifolia.* Here are reminders of histories, of men who collected the plants and perhaps grew them centuries past, fragrances of old books in calfskin and vellum.

If we are interested in the word Thymus we will find it in the Greek, associated with *incense* as one might suppose from the pungent aroma, taken into Latin for the thyme plant; it has no connection with the English word *time.*

Our *Thymus Serpyllum* is native abundantly in many parts of Europe and in central Asia as well as northern Africa. It is extensively naturalized in parts of North America, in some places in the East giving the landscape a purple tinge. It is also variable, and several binomials have been applied to the forms which are sometimes regarded as distinct species and are in gardens under separate names.

Now may we look at the specimens, as I take the cover in hand. First we see a plant from Mount Athos in Greece collected by Ballalas; then from Ingria, old district of Russia, 1860, "in locis arenosis siccis hinc inde copiosissime," which means that it was found in a dry sandy place and was very abundant; two localities in Denmark; chalk cliffs facing the sea at Freshwater in England; at Wagner Bay in the Island of Guernsey; open meadows and moors in Ranettan in Banffshire, Scotland; open field in Province Quebec; dry soil in Dorset, Vermont; three sheets from the Berkshires of Massachusetts, in meadows and low grounds; covering miles of fields and hillsides at Grand Gorge in the Catskills by myself in New York, and in Michigan introduced nearly fifty years ago; seven sheets in a botanic garden in Germany in variety, one in the botanic garden in Edinburgh, six from similar institutions in North America; specimens of my own grown from French seeds and from American seeds, others cultivated by an American nurseryman, grown in southern California, hybrid with *Thymus pulegioides* wild in Spain at about 4,000 feet elevation. Here is a sweeping fragrant journey. Here also are indisputable records of distribution and identification.

Unfortunately, not many herbaria attempt to incorporate adequate material of carefully determined cultivated plants. These plants have received far too little systematic study. Nor is there a recognized need for keeping the plants in domains and gardens true to name as there is for wild plants. Too great dependence is placed on the label. There is, to be sure, demand for registration of horticultural varieties, but that is quite another subject in a separate field. Some day cultivated plants will be recognized to be worthy of record as showing our resources in different epochs and regions. Descriptions and printed notes are not real records of species of plants. But when that day comes, some of the species will have passed from cultivation and records cannot be obtained. The best records are contemporary.

Fortunately, Linnæus made an herbarium. It is preserved by the Linnean Society, in London. It shows what was had, at the beginning, although some of his material was early destroyed. It is naturally the most important single personal plant record in the world. In cases of doubt as to what he meant by a given species, competent persons may go to the specimens themselves or have them examined by the custodians. This does not mean that he made specimens of all the species he described; some of the species are founded on descriptions and plates in previous books, and these become evidences, but of course they are not as infallible as the plants themselves.

Linnæus left instructions about his herbaria, meaning that there were two. "Let no rats or moth injure them. Let no naturalist steal a single plant. Be firm and careful to whom they are shown. Invaluable as they are, they will increase in value as time goes on." He stated that they comprised the greatest collection in the world (Jackson). "Do not sell them for less than 1,000 ducats," which would be approximately $2,300. Yet the collection became greatly damaged.

Index to the Linnean Herbarium as it is now very carefully preserved discloses 13,832 sheets. This is indeed small as compared with the hundreds of thousands, or even millions, in the leading collections in our acquisitive and recording days. These contrasts indicate the growth of knowledge in two hundred years; and one wonders what will be the astonishing treasures in two centuries to come. Perhaps ten centuries hence persons will know so much as to be confused of their knowledge.

There are those who suppose that such treasures will not need to accrue as rapidly in the coming years seeing that the world is now explored. This is a precious fallacy. We may be able to place names all over the map of the world, but this does not mean that the areas are really known. Relatively few regions, even the oldest ones, have yet been completely explored for plants; in fact, some of the oldest regions historically are among the least known. More critical exploration is disclosing overlooked species of plants in New England and New York and other territories long ago well mapped and contoured and taxed. Probably the world is not yet half really known; and I doubt whether we have collected and named one-half the kinds of plants. Vast regions of abounding vegetation are yet untraversed by the collector. New species are not discovered by airplane.

As the number of the known species of plants increases the more critical does the identification and description of new ones become. When Linnæus described and named his nine species of Cratægus (hawthorn), it was simple enough to distinguish between them, and the accounts were brief; now when we know 900 species, it is evident that greater pains must be taken to separate one from another, with closer study, more detail, greater care not to duplicate and confuse names. This increasing complexity requires the clearest records both in herbarium specimens and in literature. Moreover, the species formerly described must ever be subject to greater scrutiny, and be more and more clearly defined.

The naming of plants is increasingly much more than making an enumeration. The present-day systematist knows plants both in the field and in the herbarium; he takes into account their ranges or distribution, habitats and soils, ecological relations, variations, behaviors, and as far as possible the heredity: his problems are biological.

Moreover, the literature or written record of the subject is rapidly increasing. It is scattered in many books, proceedings, journals, separate contributions, in many parts of the world in many languages. The worker must acquire skill in bibliography and citation as well as in observation. The mere problem of keeping the names straight, clearly defined, adequately published, assumes large proportions, that the future may have less trouble with our work than we have had with that before us. It may seem a simple thing to name and describe a new species of plant, but the effort takes one far afield and away into the past.

In the seed-plants or sporophytes something like a million recognized binomials have been applied. There are other great numbers in the "flowerless plants," as the ferns and allies, mosses, fungi, liverworts, lichens, algæ, bacteria. Perhaps half or more of these names are synonyms or duplicates. Great numbers of new species are being described every year. In fact, probably there has never been such great activity as now in the founding and naming of species and natural varieties, nor ever before such painstaking and critical work. I suppose the same may be said in zoölogy. The approach to the subject has changed radically in the last quarter century. In all this excellent work the central problem is identification.

For horticulturists and botanists alike, the primary problem is not nomenclature but identification.

The usual interest in plants is associated with stature, shape, texture, color, fragrance, season, habit, habitat, tractability to cultivation, and this is correct; if to this response is added something of the life history and also a sensitive knowledge of differences, one is led into the larger beauty.

IV

RULES OF NOMENCLATURE

NOMENCLATURE means the naming of things under a system. Its root is Latin, *nomen,* name, a word also taken over into English. To plant-growers, nomenclature is likely to represent a nightmare of names. The word itself is apparently difficult, if one may judge by the different ways of mispronouncing it. The word is accented on the first syllable, with a long o: no-menclature.

Common, vernacular, English names of plants do not constitute a method. Each name is a law unto itself; it may originate without reference to any other name; it may be an old folk-name, or a chance appellation; it may be a degenerate form of another word, as "markery" is of "mercury." It may be merely a translation of a Latin binomial, as "spotted begonia" for *Begonia maculata:* these transfers, being merely verbal, are not likely to become common. Another class of cases includes Latin generic names that have become vernacular, or technical and common names that coincide, as begonia, aster, acacia, spirea, clematis, geranium, magnolia, smilax, weigela, asparagus.

Vernacular names are of all kinds and degrees of usefulness as well as of origin. Some of them are in process of becoming obsolete, and in time will be only historic. Common names represent a growing more or less changing vocabulary.

It is a fascinating quest to trace the real living vernaculars, those that have become embedded in language. They have interesting relations with habits, ideas and practices in times past. They do not constitute a connected procedure, however, and do not come within an orderly system for the naming of plants.

If a person is interested in a given vernacular name, he goes to the dictionary for its orthography, origin and meaning, not to a code of nomenclature. Perhaps he can trace it through several languages. Its root may be Anglo Saxon, Old German, Danish, French, Latin, Chinese, American Indian. Value of a common name is determined by usage rather than by priority.

For common names of plants, therefore, the reader is referred to an unabridged dictionary, particularly if he is skilled in tracing origins as given just after the entry of the word. Every word is an historical story. If the reader wants lists of English names, he will find them in the indexes to the different botanical manuals, and they are given in the text along with the Latin binomials. There are also special books devoted to the common names of plants. A book of critical value for general reference is the Dictionary of English Plant-Names, J. Britten and R. Holland, published about fifty years ago by the English Dialect Society. There are a number of smaller books in England and America. Standardized Plant Names, 1923, prepared by American Joint Committee on Horticultural Nomenclature, is replete in English names of cultivated plants, old and new. A monumental work of international character is the two-volume Dictionary of Plant Names by H. L. Gerth Van Wijk published at Haarlem in 1911 and 1916 by the Dutch

Society of Sciences, giving lists in English, French, German and Dutch. The student will find many aids if he enters the fertile field of the common names of plants.

Common names lack precision; therefore, their practical utility is limited. Sage-brush may mean several kinds of plants; soft maple means different species of maple, depending on the region; in fact, maple itself may mean Acer or Abutilon, or in Australia something different from either; huckleberry has no definite application; dogwood is one thing in North America, another thing in England, and still another in the tropics; cowslip is a swamp plant in the United States, an old garden flower in England; pine is Pinus in the northern hemisphere, Araucaria, Callitris or other things in Australia; even the familiar old word hollyhock includes two species, and the pumpkin may mean three; potato is one product in New England and another in Alabama; yam of Louisiana is a very different commodity from that of the island of Trinidad; almond is a familiar nut of commerce, or a little garden ornamental bush, or in the tropics neither one; nasturtium of horticulturists is one plant, of botanists quite another plant; examples could be multiplied indefinitely.

Botanical binomials are exact. They apply to one kind of plant, critically distinguished from all other kinds. They are employed by writers in any language. Two difficulties confront the plant-grower in respect to them: they are "hard," and they themselves are likely to change.

It is true that many of the Latin names are difficult and "big": examples are Chrysanthemum, Gladiolus, Pelargonium, Gypsophila, Hemerocallis,

Amaryllis, Hydrangea, Delphinium, Aquilegia, Narcissus, Philadelphus, Pyrethrum, Ranunculus, Dahlia, Cratægus, Coreopsis, Petunia, Sempervivum, Viburnum, Calceolaria; perhaps the toughest of the lot is Rhododendron. Curious case of preference is that of Rhododendron, which seems not to be displaced by the English name rose-bay.

It will profit us to pause yet another moment to emphasize again the fact that the Latin binomial classifies the plant as well as names it. The binomial carries relationships and leads to understanding. Common names not only avoid relationships but many of them suggest false kinships: asparagus fern is not a fern, and the name should be transposed to read fern asparagus; pineapple is neither a pine nor an apple; calla lily is not a lily nor does it even belong to the lily family; pepper-grass is not a grass; horse-chestnut has nothing to do with a chestnut; grapefruit has no relation to grapes; alligator pear, an absurd name still in use, is no kin with a pear; castaneas of commerce (Brazil-nuts) have no connection with the genus Castanea (chestnut). Recently my attention was called to a man who grew tobacco from seeds obtained from Indians; desiring to know its Latin name he looked in the indexes of books for Indian tobacco and then called his plant *Lobelia inflata,* but in fact it was a true tobacco or Nicotiana.

Difficulties in the change of names may now be considered; and this brings us to the Rules of Nomenclature, the nature of which must be apprehended before one can understand names of precision. Discussion of changes and their reasons comprises the remainder of this chapter; but the botanist thinks of them not as changes but as results of procedure: he applies the rules; if a change arises it is secondary in the process. We shall try to understand the usual methods in their simple elements.

A basic principle in nomenclature is priority of publication, although the application of this law may be modified or in certain cases withheld, under proper authority, to allow of more important gains. It is agreed to begin binomial nomenclature of higher plants with the first edition of Linnæus' Species Plantarum, 1753, with which is associated the support of the fifth edition of his Genera Plantarum, 1754.

There was not a general adoption of Linnæan binomial nomenclature immediately following the publication of Species Plantarum. Thus the famous Gardeners Dictionary of Philip Miller, begun in 1731, did not adopt Linnæus until the seventh edition, 1759, and then incompletely; the great eighth edition, 1768, is his perfected use of binomials. Nor did the binomial system have the importance for many years after Linnæus that it has assumed now, with the greater number of recognized plants and the more critical care given to identification, diagnosis, and bibliography. Modern libraries are much more complete in books and periodicals dealing with the kinds of plants, and comparisons can be made more accurately than ever before. It has become necessary to formulate precise rules to eliminate old duplications and disharmonies and to prevent them in the future.

These rules, on an international basis, are recent, and we are yet in the midst of the changes resulting from the application of the latest of them although probably past the worst of the difficulties. The varying practice of nearly two centuries is to be assorted and harmonized. Before the formulation of compre hensive and careful rules, the practices in the use of binomials were largely personal or on the pattern of prominent authorities.

Authority in botanical nomenclature proceeds from international conventions of persons pursuing science. Such conventions are congresses composed of delegates or representatives of regional or departmental scientific bodies. That is, binomial nomenclature is a problem in science.

A code was adopted by a Botanical Congress held in Paris in 1867, but it did not acquire the authority attained by more recent enactments. American systematists formulated rules late in the past century, and a Nomenclature Commission was established. This Commission at a meeting in Philadelphia in 1904 approved a set of canons. This code was radically different in principle from that of the Paris Congress. There was activity in other parts of the world. An International Botanical Congress was held in Vienna in 1905, at which a set of International Rules for Botanical Nomenclature chiefly of Vascular Plants was adopted. This formulation was based on the Paris code of 1867. The American set of principles was presented at Vienna but not adopted, whereupon the adherents declined to accept the Vienna formulation and established the American Code of Botanical Nomenclature. Other Americans accepted the International Rules. Thus it came about that in the United States there have been two codes of nomenclature for a quarter century. The two sets agree in many particulars. At the Second International Botanical Congress in 1910 at Brussels, modifications were made in the Rules, and again at the Fifth Congress at Cambridge, England, in 1930; and at the latter Congress adjustments were effected and certain of the American-code position accepted.

The relative merits of the International and American rules or codes are not under discussion here; they are naturally technical and of little interest to the general inquirer. Certain features essentially common to both may be mentioned for the purpose of explaining how binomials are made and changed, and also two provisions in which they radically differ. If phraseology is quoted it is from the International Rules, to which the writer has adhered, in part just because they are international and because he has worked with cultivated plants that are native in various regions of the world and have been described in many countries.

"Natural history can make no progress without a regular system of nomenclature, which is recognized and used by the great majority of naturalists in all countries" is the opening statement of the International Rules in the English version; and the Rules are "destined to put in order the nomenclature which the past has bequeathed to us, and to form the basis for the future."

We have already learned that the Latin appellation is in two parts, the generic name and the specific; and there may be a varietal name subordinate to the species: in *Prunus Persica* (peach) Prunus is generic and Persica specific; in *P. Persica* var. *nucipersica* (nectarine) we add a varietal name. With this basis and the principle of priority in mind, we may proceed.

Each natural group (as species) can bear in science only one valid designation, and that the oldest. When a species is moved into another genus, the first specific epithet must be retained. That is, the first species-name follows the plant into whatever genus it may be placed by different authors, unless there is some special obstacle. The peach was named *Amygdalus Persica* by Linnæus in Species Plantarum; when subsequent authors combined Amygdalus with Prunus, the peach became *Prunus Persica*. Several writers in early days brought the peach over into Prunus, as that genus was enlarged to cover the pomological stone-fruits. Apparently the earliest regular transfer was by August Johann Georg Karl Batsch in 1801, Weimar, in Beyträge und entwürfe zur pragmatischen geschichte der drey natur-reuche nach ihren verwandtschaften: Gewächsreich. Tournefort called the peach Persica (the word peach is derived from Persia, whence it was then supposed to have come) and Philip Miller in a post-Linnæan edition of his Gardeners' Dictionary adopted the name as generic, and the peach became *Persica vulgaris;* this disposition has not been accepted in recent time. The synonymy of the peach, if one prefers to keep it in Prunus, becomes:

Prunus Persica, Batsch in Beytr. und Entwürfe Pragm. Geschichte, i, 30 (1801).

Amygdalus Persica, Linn. Sp. Pl. 472 (1753).

Persica vulgaris, Mill. Gard. Dict. ed. 8 (1768).

The var. *nucipersica* must follow the peach in whatever binomial it may acquire. Of course the name *Persica* cannot be applied to any other species in

Prunus, but may be written in other genera, as it is in *Syringa persica* (Persian lilac). These two uses of the word *persica* as a specific name we shall meet again.

Frequently it happens that a species must have a new diagnosis (technical description), the original account having been found to be insufficient or even in part erroneous; or what was considered to be one species (as in the case of *Prunus japonica* mentioned on page 41) may turn out to be two or more species. These changes in definition, however, do not change the name; one must only be sure what plant was intended in the original name and diagnosis, and the name holds for that plant, even though the definition of it was imperfect. That is, a name is a name, not a description.

To determine just what plant the author meant by his name and definition, his original specimen is consulted, as we have already learned: that herbarium plant is *the type*. In case (as often with the early authors) there was no type specimen, recourse is had to a picture he may have cited; the record of nativity may aid in identifying the subject. To identify the plant intended in such cases often requires clever detective work, with good knowledge of the group to which the plant belongs and the assorting of probabilities. These subjects are full of delightful puzzles.

Good example of the misinterpretation of the name of a conspicuous tree for more than a century and a half is the case of the cottonwood of the eastern United States. One of the several kinds of poplar in eastern North America is the tacamahac or so-called balsam poplar, a narrow-topped tree with very sticky bal-

samy buds and long leaves whitish underneath, growing mostly in the northern parts; another poplar is
the cottonwood, a very broad-topped tree with little
balsam odor and very broad leaves, widely distributed. (Poplar of the lumber trade is not a poplar at all but tulip-tree or liriodendron). Linnæus in
1753 founded the species *Populus balsamifera* (balsam-bearing) with "Habitat in America septentrio-
nali" (North America). He did not describe the tree
except as he quoted phrases from earlier works, one
of the references being the full definition by Mark
Catesby in the illuminated Natural History of Carolina, Florida and the Bahama Islands, 1731–1743.
The name *P. balsamifera* was confidently applied to
the northern balsameous poplar for more than a century, yet it would be strange if Catesby meant that
species when writing of the plants and animals of
Carolina, Florida and the Bahamas. Meantime Aiton
in his Hortus Kewensis of 1789, being a catalogue of
the plants growing in the gardens at Kew near London, had described *Populus monilifera* from eastern North America; this was plainly the cottonwood,
and so the name was long applied in this country. It
was discovered, however, that Humphrey Marshall
had described the cottonwood in his Arbustrum
Americanum, the first American publication on
trees and shrubs, as early as 1785, under the name
Populus deltoide. Presuming his name to have been
a misprint, the cottonwood came later to be known
as *P. deltoides,* Marshall. It was apparent that something was wrong in the nomenclature of these poplars,
but it was only recently that the Catesby specimen preserved in the British Museum was examined and correctly identified, with the result that Sargent in 1920

authoritatively applied the Linnæan *P. balsamifera* to the cottonwood, and both *monilifera* and *deltoides* became synonyms of it. This left the northern poplar theretofore known as *balsamifera* apparently nameless; but the tireless gardener-botanist, Philip Miller, in an edition of his Gardeners Dictionary in 1768 had described that tree as *Populus tacamahacca,* adapting an Indian name; and so this balsamaceous poplar is latterly known. Other specific and several varietal names are involved in these confusions but they need not be recorded here; perhaps the reader is himself by this time confused, but this is a simple case as compared with others that might be reviewed for his benefit. Question now remains whether the name balsam poplar shall still be applied to the balsam poplar, or transferred to the cottonwood (which is *balsamifera*) or dropped altogether; this I leave to the entertainment of the reader. These changes may seem grievous to the nurseryman, but are in the interest of truth.

Now may we return to consideration of the rules, about which this chapter is more or less concerned, although the poplar case shows how rules apply themselves when identification becomes finally clear. Yet it is not amiss if we pause to examine two statements in the Linnæan account of *Solanum PseudoCapsicum,* on page thirteen. In the first sentence Linnæus speaks of sessile umbels of flowers, the umbel-like clusters being without peduncle or stalk; in the second sentence, taken from his Hortus Cliffortianus, the flowers are said to be solitary. The pictures he quotes do not show the flowers to be umbelled nor are they so in the plant on my table. Descriptions of the leaves are not harmonious. There is a specimen in the Linnean herbarium in London but I have not seen it or a photograph of it. What these differences signify I do not know nor shall I now inquire; perhaps the natural variability of the plant accounts for these statements: but these are the kinds of disagreements that must be resolved when one comes to critical study.

On page 5 of this book we discovered the nomen *Solanum Capsicastrum* on a seed-packet of Jerusalem cherry. That name is in good standing, having been published a hundred years ago in a German horticultural magazine, as native in Brazil. It is reckoned a grayish plant because of thick pubescence whereas *PseudoCapsicum* is accounted green and smooth, and there are other recorded differences. These differences seem to vanish in cultivation; it has been suggested that the garden plants may be hybrids, but this point cannot be determined by surmise. Question is, whether plants grown as Jerusalem cherry are one species or two, or whether *PseudoCapsicum* and *Capsicastrum* are really distinct. We have here again a definite problem in identification to be worked out by careful study; in the meantime and until the question is determined I know the common Jerusalem cherry as *Solanum Pseudo-Capsicum* as others have known it before me.

A binomial long applied to a plant and appearing continuously in the literature is subject to displacement if an older adequately published name is found. Example is the common greenhouse heliotrope. This is always known in horticulture as *Heliotropium peruvianum,* so named by Linnæus in the second edition of Species Plantarum, 1762. It turns out, however, that Linnæus had founded a species *H. arbor-*

escens as early as 1759 in the tenth edition of his Systema Naturæ. The two plants are the same, and *Heliotropium arborescens* comes up and *H. peruvianum* goes down into synonymy.

Whether a genus shall be divided into two or more (as Pyrus into Pyrus, Malus, Cydonia) or whether two or more genera are combined into one (as Azalea included in Rhododendron) is not a question of rules or codes. Regulations provide the procedure when segregations or combinations are to be made. Such changes depend on the judgment of the worker.

Similar remarks may be made in reference to species. Thus Regel described the honeysuckle *Lonicera Alberti* from Turkestan; Rehder thinks it is not specifically distinct from *spinosa* and makes it *Lonicera spinosa* var. *Alberti*. The Swiss botanist, the first DeCandolle, called kohlrabi *Brassica oleracea* var. *caulo-rapa;* the Italian Pasquale thought it a good distinct species and named it *Brassica caulorapa*. All these authors were within their rights.

What constitutes a species is again to be judged or decided by the person, as we have learned. No one single mark or feature determines the point. Usually the systematist relies on a combination of differences; one character, as shape of seed-pod, must be found to be associated or correlated with other characters (perhaps of flowers or leaves or habit) before he is ready to describe the plant as a separate species. The tendency is to consider the plant as a whole before deciding to call it new, in respect also to range, habitat, and field characters. More characters are available than a few years ago by which to check up on specific differences. Recently aid is provided in the chromosomes, which are bodies in the nucleus recognized at time of cell-division, revealed under microscope technique. The number of chromosomes is usually constant in each pure species, as far as investigations have proceeded. This evidence is welcomed by systematists, but to base species on chromosome character alone would not be convincing. Of course we must ever be ready for any new concept of species or genus resulting from study. At present, the work in cytology (the ology of cells) is making great headway.

Two general schools of thought are in evidence in respect to natural limits of genera, some students preferring to keep related groups together in large genera and others to segregate them under special generic names. Whether the currants and gooseberries shall be kept together in the single genus Ribes, as has been the prevailing custom until contemporaneous time, or divided into Ribes (the currants) and Grossularia (the gooseberries) rests on the choice of the investigator which again is largely determined by the theory or concept of a genus. The privilege of dividing or uniting cannot be denied. Plantsmen are likely to ask why agreement cannot be reached on such questions: yes, when we agree on politics, art, economics, religion, and all else; but the larger compensation considers it to be undesirable that all persons shall be of one mind. Yet, nevertheless, when certain systematic questions have run their course, our successors may find themselves concurring in certain opinions of secondary importance that today are troublesome.

When old species-names attach themselves to a novel genus-name, what is called a "new combination" results. If, for example, the native grapes are considered to be of two natural genera, Vitis proper, and Muscadinia comprising the muscadines, then *rotundifolia* (to which the Scuppernong belongs) leaves Vitis and makes a new combination as *Muscadinia rotundifolia*. The shifting of names from genus to genus, or from species to variety and variety to species, as may follow with different personalities and closer study, results in many novel combinations. The nomenclature expresses the facts in nature as the particular author interprets them.

To avoid disadvantageous changes in nomenclature of genera by the strict application of the principle of priority, the International Rules provide a list of generic names that must be retained in all cases. The retained or conserved names are by preference those that have come into general use in the fifty years following their publication or which have been used in monographs and similar works up to the year 1890. A long list of such *nomina conservanda* was appended to the International Rules enacted in 1905, and a smaller list was added as a result of the Congress of 1910.

It is in the *nomina conservanda* that probably the greatest differences in practice occur between the International Rules and the American Code so far as horticultural nomenclature is concerned. Thus, Zinnia is a retained name *(nomen conservandum)*, Linnæus, 1759, as against Crassina, Scepin, 1758; Carya as against Hicoria; Ardisia as against Icacorea; Shepherdia as against Lepargyrea; Desmodium as against Meibomia; Dicentra against Capnoides; Smilacina rather than Vagnera; and many more.

Not every *nomen conservandum* turns out, on investigation, to be an exact equivalent of the *nomen rejiciendum* or rejected name. A case in point is the palm name Chamædorea, Willdenow 1806, as against the rejected name Nunnezharia, Ruiz & Pavon, 1794. The Willdenovian genus is founded on a Venezuelan palm, the Ruiz-Pavonian on a Peruvian palm. If further studies should disclose marked differences between the two groups, so much so as to constitute distinct genera in the opinion of a competent investigator, it would be allowable to retain Nunnezharia for the genus of Peru but it is estopped from displacing Chamædorea.

Another series in which strict application of the priority rule is halted by the International Rules but not by the American Code is when two identical names come together to form a binomial. Catalpa is an illustration. To Linnæus this tree was *Bignonia Catalpa,* thereby preserving the American Indian name. In 1771 Scopoli separated the catalpas from the bignonias as another genus, and when Thomas Walter published his Flora Caroliniana in 1788 he made the common American species *Catalpa bignonioides* (bignonia-like). Under the strict rule of priority the earliest specific name follows the plant into whatever genus it may go and the tree becomes, in that case, *Catalpa Catalpa.* The International Rules prohibit such duplication of names, and under that procedure the name of the tree is *Catalpa bignonioides.* Subsequently John A. Warder recognized another catalpa in the eastern United States, *Catalpa speciosa.*

Similar case of duplicate names is Sassafras, which we shall soon meet again. This was *Laurus Sassafras* to Linnæus, perpetuating the vernacular name. Under the practice of the American Code the name of this tree automatically becomes *Sassafras Sassafras* if separated from Laurus in the genus Sassafras; under the International Rules another name must come up. Other examples are *Malus Malus,* apple, if taken out of the genus Pyrus; *Citrullus Citrullus,* watermelon; *Lagenaria Lagenaria,* white-flowered or sugar-trough gourd, if retained in the genus Lagenaria; *Barbarea Barbarea,* winter-cress; *Vitis-Idæa Vitis-Idæa,* mountain cranberry, when separated from Vaccinium.

Publication of a new species is in a scientific journal or proceedings or authoritative book or contribution available to the public. Communication of new names at a public meeting, or the placing of names in collections or gardens open to the public, or at exhibitions, do not constitute publication, as allowed by the regulations and accepted by botanists. The International Rules require that a diagnosis, at the time of original publication, shall be in Latin, that it may be equally understandable by competent persons in all lands; this article was reaffirmed at the Fifth Congress, 1930, in England. Latin to the systematist, as to many others, is a living language; it may be very different from the classical language, however, in its vocabulary.

A new name not associated with a diagnosis or description is a *nomen nudum* (sometimes abbreviated as *nom. nud.*) or naked name, and has no standing; sometimes in lists the entry *nomen* indicates a name only and therefore not tenable. Many names long more or less current in lists and catalogues and journals must be discarded for this reason; in that case, the next name in succession of date, regularly published and not otherwise disbarred, must be adopted.

Sassafras was named *Laurus variifolia* by Salisbury in 1796, and this name has been brought over into Sassafras as *S. variifolium* (the Linnæan specific name *Sassafras* making a tautological binomial) ; but Salisbury's name is a *nomen nudum* and therefore does not count. The later name *Sassafras officinale,* 1831, is the next name in order, not barred by the International Rules. Pineapple was named *Bromelia comosa* by one of Linnæus' students in 1754 and the adjective has been brought into the present genus Ananas, but the nom is a *nomen nudum* and the much later name *Ananas sativus* of Schultes, 1830, is current.

Many floating *nomina nuda* are in horticultural literature. They originate as names on exhibition specimens, in reports of meetings, in trade lists, and become current; but as they have never been published they cannot be identified as of a given date. Thus the plant known in the United States as Boston ivy and in England as Japanese ivy was long called *Ampelopsis Veitchii;* but that is an unpublished trade binomial, and must become a synonym of *A. tricuspidata* even though we know in fact what plant was intended. The synonym is recorded as *A. Veitchii,* Hort., that is, of horticulturists or gardens. The old genus Ampelopsis was not homogeneous, and it has now been divided, as we define the categories more exactly; Boston ivy becomes *Parthenocissus tricuspidata.*

Repeatedly has it been said or indicated in this writing that names of species and botanical varieties once regularly published cannot be changed, not even by the authors of them. In the language of the Rules, no one is authorized to reject, change or modify a name because it is badly chosen, or disagreeable, or another is preferable or better known. Of course the name may not be adopted by subsequent writers, but its dismissal would be on other grounds and its form would not be changed.

Moreover, the original spelling of a name must be retained, except in cases of manifest typographical error. One may not correct them because they are etymologically incorrect; these names are technical terms. The case of Penstemon is in point. It is commonly written Pentstemon, but the earliest post-Linnæan form is Penstemon, and this spelling may be favored. Linnæus described these plants, such as he knew, under Chelone. It is said that Penstemon is linguistically incorrect, inasmuch as the name means "five stamens" and *Pent-* must be the first element; so is Pentstemon inexact; the proper form in etymology is Pentastemon, and this spelling has been revived in recent time. The simplest way is to follow rules in this case, and go back to the earliest form. Endless names would have to be changed if we tried to correct them all on the basis of linguistic form; and even then in many cases the doctors would not agree.

One of the most troublesome of the nomenclature regulations in respect to horticultural practice is the so-called "homonym rule." A species-homonym, in botanical usage, is an earlier use of the same name in the given genus. Thus, we have had before us the case of *Prunus glandulosa* of Torrey, applied to the wild Texan peach; the name *glandulosa* is a duplicate in Prunus of the earlier *P. glandulosa* of Thunberg and therefore cannot be employed in the genus as Torrey proposed. "Two species of the same genus cannot bear the same specific name" in the language of the Rules.

Nevertheless, to avoid unnecessary changes the International Rules as first adopted provided that a name need not be rejected "because of the existence of an earlier homonym which is universally regarded as non-valid," that is, dead and buried, improperly made or published, or otherwise out of use. The American Code, however, allows no exceptions: "A name is rejected when preoccupied (homonym)," and a specific homonym is defined as a name that has been published for another species under the same generic name. This provision has now been incorporated in the International Rules.

The difficulty in the operation of this regulation in horticultural subjects is not alone the fact that many well-settled names may be upset because an older but perhaps unused name may be discovered to have been employed in the genus, but because new combinations may have to be made in names of any number of cultigens with the shift in species-names, thereby complicating citations and literature without appreciable gains. The perplexing case of the Douglas fir is an example. Cases of this kind lend weight to the horticultural demand that certain names, as of important plants, be accepted and standardized by agreement and thereafter not be subject to change.

If the indication of the binomial is to be accurate

and complete, and in order to verify the date, it is necessary to quote the author who first published the name: *Parthenocissus quinquefolia,* Planch. (Virginia creeper) means that the combination of these two words to represent a particular plant is on the authority of Planchon. The addition of the authority is a form of book-keeping.

It was not always so. Linnæus did not cite authorities for binomials; his references were to literature. So with Willdenow, his editor who extended Species Plantarum into many volumes as a fourth edition (Linnæus made three). Thus *Monarda fistulosa* (common horse-balm of eastern North America) is not accredited to Linnæus although named by him. As phytography (the description of plants) became more exact and the literature expanded, it was necessary to keep closer track. Authorities came to be quoted with every binomial in all technical or floristic work.

This book-keeping has now gone a step farther. Virginia creeper, for example, was first described by Linnæus as *Hedera quinquefolia* (five-leaved). To give clue to both events, the original publication and the transfer to another genus, it is now customary to write *Parthenocissus quinquefolia* (Linn.) Planch. The name in parentheses, in case of such double citation, is the author of first or original publication. The International Rules allow such double citation: "the original author can be cited only in parenthesis"; the American Code is more mandatory: "the name of the original author should appear in parentheses."

This author-citation is of course essential in technical floristic works and similar writings; but the general public, horticulturists, nature-lovers, should not be asked to remember such citations although they ought to know what they signify when names are to be traced. The assumption that the authority is undetachable has led to the pedantry of carrying it in popular writings, even to the extent of the double citation, where the additional element becomes cumbersome, is of no significance to the reader, and may introduce a confusing distraction. It is necessary to endeavor to make binomials attractive in general writings, that their value may be more widely recognized; and this requires simplicity of presentation.

The whole subject of botanical and horticultural nomenclature is therefore well in hand in international organizations and in societies representing particular classes of plants. Standing committees of the International Botanical Congress carry the subject in the interregnums. The present organization on nomenclature comprises an executive committee of seven members, editorial of four members, general committee representing sixty-one countries and certain ex-officio members, eight special committees on the main subdivisions of the plant world.

If any cultivator has had the patience to follow this book to the present point it is undoubtedly because he has hoped for at least a paragraph about horticultural nomenclature; and we now come to this subject. It is evident that cultivated plants cannot be separated from wild species in their nomenclature seeing that they were once wild and probably are still wild somewhere. Moreover, the nomenclature of a genus cannot be divided as between cultivated and feral subjects. For example, the great genus Astragalus, with several hundred species in the northern hemisphere and many of them ornamental, is barely represented in cultivation; naturally there could not be a separate system of naming for these fortunate few. Again, rules of nomenclature must regularize the naming of new species as they are discovered, many of which are sooner or later brought into cultivation; and the names of all species, old and new, must follow the regulations.

There are horticultural varieties of species, however, and hybrids, that may not be covered by the regular rules for binomials. Both the International Rules and the American Code provide for the naming of hybrids, and the former carry a general statement on the names of "forms and half-breeds."

The Second International Botanical Congress was held in Brussels, Belgium, in 1910. At that time a subsection considered the subject of horticultural nomenclature, representing the Royal Horticultural Society of England and other similar bodies. A set of Rules of Horticultural Nomenclature was adopted by the Congress, consisting of sixteen articles. Article I provides that horticultural nomenclature is based on the rules of botanical nomenclature adopted by the Vienna Congress of 1905 "so far as they apply to names of species and groups of a higher order," but the Congress adopted modifications and additions for horticultural varieties, and hybrids of cultivated plants. Omitting the regulations on hybrids, the following declarations may be briefly noted: In naming horticultural varieties the complete name of the species to which they belong should be given; Latin should not be employed in names unless the character of the plant is expressed in such name, as *nanus, fastigiatus,* and the use of Latin proper names is proscribed;

names of horticultural varieties must be printed in Roman letters. When vernacular names are transferred to other languages they must not be translated. Varietal names should be a single word and not more than three words. Publication of a description of a variety in a dated catalogue is valid, but the mention of a variety without description in a catalogue, or in the report of an exhibition, is not valid publication even if a figure is given. It is desirable that descriptions of new varieties published in horticultural catalogues should also be published in periodical horticultural papers. In order to be valid, the description of a new variety or of a new hybrid must be drawn up either in German, English, French, Italian, or Latin.

Horticultural nomenclature on an allied basis is now in the hands of a permanent committee appointed by the last two International Horticultural Congresses. This committee will pass on scientific botanical names and also on the vernacular names of horticultural varieties. A preliminary list of adopted generic names has been issued. The report of the Ninth International Horticultural Congress, held in England in 1930, has been published by the Royal Horticultural Society.

In North America important rules of horticultural nomenclature have been adopted by organizations for the particular class of plants in which they are interested. Prominent codes of long standing are those of the American Pomological Society for fruits, and of the Committee on Nomenclature of the Association of American Colleges and Experiment Stations for kitchen-garden vegetables. Agencies for registration of varieties also provide for protecting the name although perhaps not for constituting it.

Most important part of the rules adopted at **Brussels** in 1910 is in the first article, specifying that in naming horticultural varieties the complete name of the species to which they belong should be given. This implies discrimination between species and varieties, an attitude none too common, but unless one has this primary knowledge the subject of nomenclature cannot be understood. It is common practice to omit the specific name altogether and to place the varietal name directly against the generic name. Before me is a catalogue listing *Prunus grandiflora*. There is no such plant as that. It is, I suppose, a form of one of the recognized species of Prunus. Here are species and Latin-named varieties of azaleas all listed as of equal rank, with no information to the reader as to natural relationships. Such cases may not be the fault of the nurseryman who propagates and sells the stock; he in turn takes the plants with the names under which they come to him; but somewhere along the line names have been loosely or inaccurately made or applied, very likely at or near the point or origin.

Horticulturists complain of the difficulties in botanical nomenclature: very well; here is one of the reasons for the confusions. As long as this practice is continued of treating varieties as if they were species there is no use in asking for a stabilized nomenclature. This problem of the proper usage in Latin-named cultivated varieties should receive careful consideration by competent horticultural societies. These varieties may be much more important to the grower than the type of the species itself, but nomenclature should not be confused by them.

Let it be explained, however, that there are marginal cases in which the omission of the specific name is allowable; they are those in which there has been variable botanical practice. Thus the beautiful florist orchid *Cattleya gigas* was regularly described as a species by Linden and André in 1873; others have referred it to *C. labiata* as one of the variants of a polymorphous species, along with *C. Warscewiczii, C. Luddemanniana* and others. Other authorities prefer to keep *Warscewiczii* distinct as a species and to refer *gigas* to it as a variety. Orchid growers retain the original name *C. gigas,* for which they have perhaps ample authority even though the plant may have the characteristics of a variety rather than sufficient marks to constitute a species. Such cases are few enough to be exceptional, and they are defended in the fact that they follow a recorded procedure so that confusion does not result.

Another class of horticultural cases may be described. These examples are properly part of the discussion of identification in the preceding chapter, but they are brought up here to show that rules of nomenclature may give us no help. I am fond of pinks. From seeds and roots I have grown *Dianthus caucasicus, D. cruentus, D. erythrocoleus, D. graniticus, D. procumbens, D. Sternbergii, D. strictus,* all of which turned out to be maiden pink, *Dianthus deltoides.* I like the maiden pink and was glad to have the testimony of so many names.

Now it happens that all these Dianthus names represent supposed separate species; then how? Maiden pink is a hardy persistent creeper; very likely patches of pinks of several species were grown in nurseries side by side; one by one they died out, except the maiden and this one covered the territory: the stake labels remained. I have had similar experience with

Thymus, Veronica, Campanula, Sedum. I have recommended to plantsmen that they do not grow their stock plants of the same genus together or side by side.

Horticulturists are prone to over-estimate the importance or at least the terrors of the nomenclature question, or to expect too much from rules and codes. Many of our most difficult problems with the names of plants are not clarified by regulations, as we have learned in the preceding chapter. Let us consider other cases, lest we forget.

In all the great groups of cultivated plants we are troubled by the multitude of names of horticultural varieties. There are thousands and hundreds of varieties of apples and peaches and pears, of potatoes, onions, of dahlias, sweet peas, chrysanthemums, strawberries. Many of the names are duplicates; that is, the same or essentially identical variety may receive two or more names, perhaps a dozen. It is desirable to eliminate these duplications and reduce the names in authoritative lists. It is by investigation that these duplications are discovered, sometimes by extended tests in trial-grounds. With the accumulation of evidence, the duplicate names may be discarded; rules may be required to facilitate the editing of names, but the problem is identification.

In every new edition of a book dealing with the flora of a region or with a group of plants, certain changes in names appear. Mostly these changes are results of new evidences on the identity of the plants. I may cite an example that has recently been in the horticultural press. It is the case of the so-called "Chinese evergreen" or "Chinese water plant" introduced obscurely within recent years and now employed as an indestructible window-plant. It had not bloomed for some time after introduction, and although plainly an aroid, its genus was not recognized. It was entered in Hortus as *Aglaonema simplex*. Now the plant has bloomed in different places and it is determined that the species is *Aglaonema modestum*. This is not a problem in nomenclature or a question of rules. Both *A. simplex* and *A. modestum* are approved names for two species in Malaya, but, so far as we know, only *A. modestum* is grown in this country; it is a case of mis-determination, to be corrected in a forthcoming edition. Such cases are frequently arising in all classes of plants. Were it not so, the situation would be evidence that we are not alert.

Demanding an invariable nomenclature in plants, we may yet habitually accept the opposite in other fields or subjects. The nomenclature of human beings is particularly troublesome, with the change of names by marriage and remarriage, by pen-names and stage-names, combinations of paternal and maternal surnames, emphasis of middle names, by the bestowal of titled ranks, and the varying practices with different peoples; yet we make no complaint. The nomenclature as well as the terminology of all the sciences, as well as of arts and industries, has changed and extended radically within the span of the older of us. It must be a very good world in which so much novelty constantly appears.

Even with all the congresses, rules and committees the names of plants cannot be finished. They may be

regularized. Many of the changes in names lie beyond all rules and codes of nomenclature. Linnæus
founded the genus Pyrus for the pear, apple, quince
and other pomes. Philip Miller separated the apples
in the genus Malus, a segregation long disregarded
but now accepted by many students. Names follow
these separations. In the one case the native eastern
crab-apple is *Pyrus coronaria;* in the other it is *Malus
coronaria.* It is a difference of opinion as to what constitutes a genus in a particular case; this opinion rests
on study of the plants not on study of rules. No one or
no body can or should attempt to control such opinions,
founded on research. These are biological problems,
and scientific findings must have free interpretation.

When I began the study of plants, there were two
species of Antennaria, interesting little everlastings
of the fields now sometimes colonized as ground-
cover, *A. margaritacea* and *A. plantaginifolia.* Soon
the former was placed in Anaphalis, where it ought
to be. Now about a dozen species of true Antennarias
are known in the same territory. We have learned
much since then, we have explored the country; we
have become more exact in field work as in everything
else. All these Antennarias must have names, and rules
cannot prevent it. New knowledge must be recorded.

Field work is more extensive, more thorough, more
critical, and therefore more useful and delightful, than
ever before. We are seeing things long overlooked.
We re-define species supposed to be well understood.
We are more and more convinced that we understand
nothing in the sense of finality. Our successors will
disagree with many of our findings; we wish them
well. The world, we have found, is very young so far
as our knowledge of it is concerned.

These field studies are biological problems of the highest interest and importance. In them are physiography, ecology, and many things we cannot name. Every group of plants needs re-study at least every twenty-five years. What we now know about the hawthorns, wild blackberries, panic-grasses, dogbanes, pondweeds, irises, bears little resemblance to what we thought we knew twenty-five years ago.

Nature has no strait-jackets. Plants are plastic. They vary, often for reasons we do not know. We could not have a stable invariable nomenclature even for buttons unless for all time we could control the materials from which they are made, the machines that make them, the persons who want them. Those who look for a changeless nomenclature should change their notions quickly to avoid disappointment.

The reader should now be prepared, if he is still awake, to be told that systematists are not jugglers of names. They are as much entitled to their findings as are chemists, physicists, economists, archeologists, or philosophers. The names follow investigation. They follow under regular rules of procedure, but the necessity of them precedes the rules.

This does not mean that the binomials are confused. It is impossible to have general confusion when the work proceeds under regulation and all the steps and processes are recorded. Some cases are intricate and complex, and there may be difference of opinion on the application of even the most precise rules, in harmonizing the historical complications of centuries and in making contact with very variable elements in nature. The application of any workable rules of nomenclature is technical and can be fully understood only by years of experience. This means that the application must be left to competent persons. The regulations provide an interesting system, with which it is a pleasure to work.

It is desirable, naturally, to have botanical names as uniform and understandable as possible; that is what rules and codes attempt to accomplish. Probably the regulatory changes will not be as great in the coming years as in the last twenty-five. To put into effect the systems adopted in 1904 and 1905 necessarily required many rapid changes in the interest of ultimate harmony. Changes due to biological study will necessarily continue, to record discoveries and progress. There should be no fear of change: it is stimulating.

If we do not acquire uniformity, we nevertheless arrive at orderliness in nomenclature, with recognized and recorded practices; perhaps this is as far as we should expect to go. Nomenclature is not a subject separate by itself, but a concommitant of the study of plants.

While binomials should follow regularly accepted authoritative procedure, it is nevertheless legitimate for any responsible body of persons to adopt a set of such nomials to be employed for trade purposes for a specified period. This is standardization rather than nomenclature.

Zealous growing of plants requires knowledge of them; also knowledge of weather, soils, seeds, manual practices, fertilizers, diseases, pests. It is satisfaction in itself to acquire this education by study and experience. It is good to be up-to-date. The horticulturist should also wish to understand names. It is not difficult to acquire a sufficient practical and reading knowledge to make the subject interesting. This is the way to overcome fear of them; they will always be present. Forms of life begin to arrange themselves. Thereby garden and field assume fresh significance.

V

A FEW MORE

A FEW more examples of how plants get their names may be presented, as a matter of interest to the horti-culturist. The binomial of every cultivated plant is in itself a history.

To the plant-grower busy with his operations, in-tent on the care of beautiful plants under glass, in-terested in the great fields of nursery stock grown to perfection of uniformity, to the plant merchant in the market, the superintendent of parks and estates, to the home gardener concerned in making the most of a small area, to the fruit-grower, florist, all the dis-cussion in this book may seem to be foreign, trouble-some and tiresome, of no relation to living things. What to him is a musty old tome in cryptic Latin, or a set of rules about names, or complicated puzzles, or dead plants secreted in cabinets?

So be it: let us consider further cases. But in ad-vance it may be noted that the naming of horticultural plants is only a very small part of the nomenclature problem. Probably less than one per cent of the known plants in the world are in cultivation at any moment outside botanic gardens, and a relatively small number is domesticated. The main subsistence plants of the world are probably not more than about one hundred species. We have really made limited use of the possibilities of the vegetable world.

Moreover, changes so troublesome to us in 1932

will not bother those of 1950 or later because they will have been accepted; our successors may have other difficulties, worse than names.

Potato has been *Solanum tuberosum* from the first, 1753. Linnæus cites *Solanum tuberosum esculentum* of Caspar Bauhin, who in his Prodromus of 1620 gives an extended account of the plant, and a picture; the engraving is reproduced overleaf. There are further citations in Hortus Cliffortianus to which Linnæus refers; one of them is to Robert Morison, Plantarum Historia Universalis Oxoniensis, in the third volume published at Oxford in 1699 after the author's death. Morison's picture of *Solanum tuberosum esculentum* is reproduced on page 87; resemblance will be noted to the habit sketch from Bauhin. In his own characterization of the plant in Species Plantarum, Linnæus makes no mention of tuber-bearing; and probably the utility was not then great if one may judge by the curious tubers shown in the pictures; he was not discussing the uses of the plant. He gives the nativity as Peru; long before this time, 1613, Besler had described the plant as *Papas Peruanorum*.

The aboriginal word variously rendered *papas, batata, batatas,* and adopted into modern languages, appears twice also in binomials of tuberiferous plants. Once by Linnæus it is used as *Convolvulus Batatas,* supposed to be native in India. Probably Linnæus did not know the flowers. Later by Poiret it was placed in the Linnæan genus Ipomœa as *I. Batatas,* as we know it today; in the vernacular it is sweet potato. In 1854 Decaisne described a yam as *Dioscorea Batatas,* a Chinese twiner producing large under-

ground tubers, in the North known only as an ornamental under the name cinnamon-vine.

Solanum tuberosum esculentum.

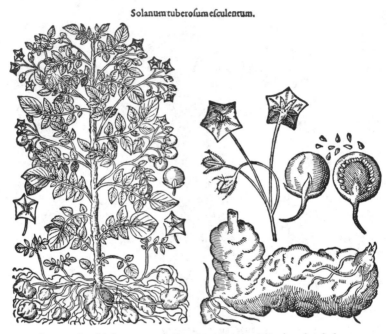

Caspar Bauhin's potato. 1620, the year the Pilgrims landed at
Plymouth Rock.

This history is simple enough, but the difficulties
with the nomenclature of common potato are just
beginning and from a different direction: is the
potato one species or several? We are more and more
in doubt as to the identity of even the potato, and the
problem is now being studied from the point of
view of genetics and plant pathology as well as
geographic distribution in the wild. Explorers have
recently been in the field in tropics to discover
original forms of tuber-bearing solanums.

As indication of departures in the understanding

of long-domesticated plants we may cite the work of the Russians, Juzepczuk and Bukasov, 1929, in

Robert Morison's potato. 1699. In the super-scription is the reference to C. B. P. (Caspar Bauhin's Pinax) and to the Papas Peruanorum of Hortus Eystettensis.

separating *Solanum tuberosum* into the following species as a result of chromosome studies:

Solanum tenuifilamentum
Solanum Juzepczukii
Solanum stenotomum

Solanum phureja
Solanum ahanhuiri
Solanum goniocalyx
Solanum Rybinii

Most of the ancient staple crop-plants have a simple nomenclatorial history inasmuch as they were named by Linnæus and there has been relatively little inquisitiveness about them until recent time: examples are rice, wheat, oats, rye, maize, banana, cabbage, lettuce, alfalfa, date, coconut, wine grape, pea, onion, pear. Recently, however, under the closer study of variation, distribution, and heredity doubt is beginning to be expressed as to whether some of the staple crops are really one species, whether the accustomed name may represent only one original kind. Is *Triticum æstivum (vulgare),* wheat, one species, or is it thirteen as Vavilov and his associates recently suggest on the basis of genetic studies? If the latter, then of course the binomials are multiplied. The nomenclature of many of the anciently cultivated plants is likely to be upset on biological grounds; this introduces a new complication. Indeed, it is from this direction that we are to look for some of the greatest upsets in nomenclature.

Some of the groups of cultivated plants have been so much hybridized that the current kinds can hardly be referred to regular original species. Thus Voss proposed the name *Begonia tuberhybrida* for the confused and crossed tuberous begonias, and comparable names in other groups. The prevailing gladioli (this word is preferably gladì-olus, gladì-oli) are so mixed that I have felt obliged to propose the name *Gladiolus hortulanus* for them. In cannas I have thought it

88

necessary to make two names, *Canna generalis* for the common flowering kinds and *C. orchiodes* for the orchid-flowered group. Such class binomials are comparable with *Pyrus Malus* for the apple (which may comprise few or several original stems), *Triticum æstivum* for the wheats, *Cucurbita Melo* for the melons, *Pyrus Lecontei* of Rehder for the Le Conte-Kieffer class of pears, *Ceanothus Veitchianus* of Hooker for a race of garden hybrids, and any number more.

It is not generally realized that the botany of the main cultivated plants is little understood, or that more critical study may make considerable changes in the names applied to them. If we do not know what sugar-cane is, we may have novelty in names when we find out.

Many of the ancient domesticated plants are cultigens,—known only in cultivation, not yet recognized anywhere as native or indigenous. The common garden and field bean, *Phaseolus vulgaris,* is an example; also Indian corn, banana, oats, rye, sweet potato, date. We are not certain of the native place of the coconut. Another example is the florist chrysanthemum, and we may pause a moment with its name inasmuch as names suggest natural histories.

The garden and greenhouse chrysanthemum is a relatively recent plant in the western world, having been introduced to Europe from the Orient in the latter part of century before last; it was figured in Botanical Magazine, England, in 1796, a purple flower much like the strain of hardy border chrysanthemum of the present day. It is to be noted that the plant had been ameliorated by long cultivation when discovered in the seaports of China and Japan by Europeans. It is probably of ancient domestication in the Orient.

An undeveloped Asian chrysanthemum was known to the herbalists, and it was named *Chrysanthemum indicum* by Linnæus, although not native of India as we understand that geographic term today. For the most part, at least until recent years, the florist chrysanthemum has been known as *C. indicum*. Much has been written on its history and origin, gleaned from many printed references. When looking for the wild chrysanthemum back in China some years ago I became convinced that we do not know enough about the native species of that part of the world to enable us to make positive statements on origins; it is a biological rather than historical problem. "It is my conviction," I wrote at the time, "that we should not speculate further on this subject until the wild forms in China are well collected, over a wide range, and are assembled for study."

Before that time, in 1914, I had proposed a class- or species-name for the cultivated plant, *Chrysanthemum hortorum,* as we have a collective binomial *(Prunus domestica)* for the common orchard plum, also a cultigen, and likewise *Citrus sinensis* for the cultivated sweet orange. Two earlier names had been applied to the oriental chrysanthemums, *C. morifolium* (morus- or mulberry-leaved), 1792, and *C. sinense,* 1823; these were supposed to represent the wild or native form of chrysanthemum and to be, therefore, synonyms of *C. indicum* or else *morifolium* a separate native species and one of the parents (with *indicum*) of the modern garden races; but on reviewing the circumstances of publication it was discovered that they also are names founded on the introduced cultivated chrysanthemum and not on an indigenous species. Therefore, the first cultigen name stands, and *Chrysanthemum morifolium* of Ramatuelle is the tenable name for the domestic group, as far as we yet know, with *C. sinense* of Sabine and *C. hortorum* of Bailey as synonyms.

The name *C. indicum* stands separately, by itself, for a wild oriental chrysanthemum. This situation I explained in print about ten years ago. When we have finally uncovered the botanical origin of the florist chrysanthemum we may give it the name of the prototype (if the name is an old one), or if the cultigen proves to be the result of amalgamation of two or more species, the name *morifolium* will probably still hold. Meanwhile, we shall continue to grow chrysanthemums.

Good examples of name changing due to confusion are the pomological blackberries. These fruits have come into cultivation from native wild berries within a century before our eyes, and yet until recently we have not been able to refer the kinds to their species with any degree of certainty, and even now we are not positive of any number of them. The pomological varieties have been given names, to be sure, as Lawton, Kittatinny, Snyder, Taylor, Lucretia, but what species they came from is another matter.

The difficulty in this case is twofold: no records were made of the varieties in the beginning, in the way of herbarium specimens; knowledge of the wild blackberries has lain in utter confusion. Many of the major groups of native plants have been singularly confused, but probably the blackberries are the worst. Comparable case is Cratægus or hawthorns, which we have already had before us, and there are any number more, as Rosa, Viola, Agrimonia, Brassica, Amelanchier, and others mentioned on preceding pages. When one of the familiar groups is worked over with additional material and extensive field experience and new binomials are introduced, those who know nothing about it are shocked; yet if the chemists discover new elements or the astronomers new stars or the physicists new explanations they are applauded. I remember how certain persons were disturbed when Coulter and Rose, more than forty years ago, presumed to describe new species of umbellifers and to change generic names, and thereby upset the orderly arrangements we had known; yet they did not invent those plants out of mischief but found them in nature and it was not their fault.

We are beginning to find order in the native blackberries and we know that there are many more species than had been supposed. Rubus (blackberries and raspberries) is one of the big genera of the North American flora, comparable in respect of size with Carex, Panicum, Aster, Solidago, Eleocharis, Artemisia, Quercus, Ranunculus; the fact that we have not known it does not change the situation. Still in existence are a good number of important botanical specimens of pomological varieties, forty or more years old, to give us a clue to the actual origins when we are competent to discuss them. This will make no difference in the value of the present cultivated varieties but it will give us information, make us aware of our natural resources, and it should enable us to plan future breeding with some hope of success.

Roses are naturally confused in botanical nomenclature because the usual cultivated ones are so much hybridized. Often it is difficult to make out the original species. Horticulturists have classified the kinds into groups without particular reference to the species involved in them, as Teas, Hybrid Teas, Hybrid Perpetuals, Noisettes, Ramblers, Bourbons, Sweetbriars. Rosa is a difficult genus in nature because variable and widely distributed.

It would not profit us to endeavor to reconstruct the origin of roses in a writing like the present: we need an authoritative attractively written book on that subject, in new phraseology, wrought not from the ordinary point of view of history but from that of botanical development; such a book would do much to clarify our ideas about roses.

As in Gladiolus and Canna, collective binomials have been developed to designate main floricultural groups: *Rosa borboniana,* the bourbon roses, including the hybrid perpetual class; *R. dilecta,* the hybrid teas to which American Beauty belongs; *R. Noisettiana,* the noisettes; *R. polyantha,* the polyantha roses; *R. damascena,* damask roses; *R. Bruantii,* Bruant roses; *R. Barbierana,* Crimson ramblers; *R. alba,* attar roses; *R. Penzanceana,* the Penzance briarroses. These names represent roses unknown in a natural or native state, being direct products of domestication or hybrids or mutants of long standing. Other roses, well recognized as parents of cultivated races and known also as wild species, are *R. odorata,* of China, tea rose; *R. chinensis,* also of China, China or Bengal rose; *R. multiflora,* of Japan, multiflora roses, and the Chinese representative of it, very distinct as seen in the wild, *R. cathayensis; R. gallica,* Europe and western Asia, French rose; *R. centifolia,* Caucasus, cabbage rose; *R. Banksiæ,* banksia roses; and a good number more. For the most part, roses are not known to cultivators in terms of Latin binomials, so that the nomenclature resolves itself into the standardizing of vernacular names of horticultural varieties.

"What's in a name?" cries Juliet; "that which we call a rose by any other word would smell as sweet." Yet Shakespeare might admit that a rose is not less sweet because we know its name. In this later day we wish to make sure that a rose is really a rose: the "bridal rose" of gardens in warm regions and of old greenhouses is a Rubus; I have eaten the raspberry-like fruits of the single-flowered form of it.

Gloxinias are not gloxinias. How this comes about is quickly told. It is a good example of the way such things come to be.

The genus Gloxinia begins with Charles Louis L'Héritier de Brutelle who lived from 1746 to 1800 and who wrote notable systematic works. The genus was founded in 1784 on a Brazilian plant, which L'Héritier named *Gloxinia maculata,* the generic title being in compliment to Benjamin Peter Gloxin, physician and botanical writer of Colmar near Strassburg. I have not seen this plant in cultivation in the United States, but it is in evidence in the American tropics. It is an attractive perennial rhizomatous herb one foot and more tall but of spreading habit, with large thick heart-shaped ornamental leaves more or less tinged violet on the upper surface and of lighter color underneath, the erect stalks bearing several or many pubescent deeply bell-shaped lilac flowers more than one inch long and accompanied by large leaf-like bracts.

Early in the past century another plant was introduced from Brazil. It was named *Gloxinia speciosa* and also pictured by Conrad Loddiges and Sons in the Botanical Cabinet, 1817, with the statement that "this most splendid subject has lately been introduced from South America, a country richly abounding in the most beautiful productions, which unhappily have been till now mostly shut out of the civilized world. The time, however, seems approaching when these treasures will be freely diffused. If the oppressions which men exercised upon each other during the dark ages of ignorance and barbarity were once to cease, all would feel the advantages, and enjoy the comforts of amicable commerce; that source of such incalculable benefits to nations." Now we have the commercial age, and still do we look for something better.

This *Gloxinia speciosa* soon attracted attention, and it was shown in colored plates. The original strain is pictured with drooping flowers whereas the florists plant produced from it has erect or ascending flowers; it would be interesting to trace the development of the present race of gloxinias step by step through a century, to disclose the art of breeding and to determine whether hybridity has entered into it, as has been said. The present gloxinia, in habit and foliage and bloom, is one of the choicest of pot plants.

In 1825 in a French periodical Christian Godefroy Nees von Esenbeck of the University of Bonn established the genus Sinningia on a Brazilian plant introduced by M. Heller, inspector of the Royal Garden of Wurzbourg, and cultivated in the University garden at Bonn; the plant was named *Sinningia Helleri*. Generic name is in compliment to William Sinning, gardener to the University of Bonn.

In 1848 J. Decaisne founded the genus Ligeria in the great French horticultural journal, Revue Horticole, naming it in honor of Louis Liger, author of many works on agriculture and gardening. He brought Loddiges' name *Gloxinia speciosa* into his new genus.

When Joannes Hanstein of Berlin wrote the Gesneria family for Martius' monumental Flora Brasiliensis, he retained the three genera and followed Decaisne in placing the Loddiges plant in Ligeria as *L. speciosa*. This account is dated on the title-page 1857–1864.

When a new study of the gesneriads was made for Bentham and Hooker's Genera Plantarum, and published in 1873, Ligeria was merged with Sinningia as not being sufficiently distinct; and our plant later became *Sinningia speciosa,* where it now rests.

These various events may be shown in a formal way as follows:

Sinningia speciosa, Nicholson, Ill. Dict. Gard. iv, 437 (1888).

Gloxinia speciosa, Loddiges, Bot. Cab. i, 28 (1817).

Ligeria speciosa, Decaisne, Rev. Hort. ser. 3, ii, 464 (1848).

Other events are in the records dealing with varieties and plants supposed to have contributed to hybridization, but they do not directly involve the name of the glasshouse gloxinia.

In the foregoing disposition of the case there is no guaranty, however. Any competent investigator, with more material before him, perhaps exacter methods, and a world range of attack, may arrive at other conclusions on the limits of genera and species; and we have learned that names follow identification.

Of Gloxinia there are about a half-dozen species native Mexico to Brazil and Peru. Apparently only *G. maculata* is much known in cultivation. Of Sinningia there are about twenty species, but only *S. speciosa* appears to be in general cultivation. Both genera yield subjects of high interest and ornamental value, and several of the species are in horticultural literature. Plants of this kind require careful glasshouse handling and the skill of the trained gardener, and they are not in evidence in the present day of standardization, much to our loss.

Before leaving L'Héritier we ought to know that it was he who broke up the Linnæan inharmonious genus Geranium, separating Pelargonium and Erodium. This was accomplished in his striking folio, Geranologia, published in Paris in 1787–8. The horticultural interest in geraniums was great in the early part of the past century, as witness the many early colored plates in the periodicals, the two-volume treatment by Henry C. Andrews, 1805, and the six-volume work of Robert Sweet, 1820–30, both published in London. A century passes; open on my table is the heavy technical volume by Knuth on the Geraniaceæ published in Leipzig in 1912. But even though the florist geraniums are pelargoniums, they are still known as geraniums, the old name persisting from the time of Linnæus and before, the same as Sinningia is still popularly known only as gloxinia.

Words are only sounds quickly emitted and then lost, yet do they persist century by century.

Sometimes amaryllis is amaryllis, but more often it is not; and thereby hangs another tale.

The amaryllis case, like that of the gloxinia, devolves on the interpretation of genera. At first the genus Amaryllis was conceived broadly, in the days when geography was more or less indefinite and sources of cultivated plants were little understood. At present the genus is interpreted as consisting of a single species (with marked varieties and races) native in the coast region of Cape Province, South Africa, and the name of the species is *Amaryllis Belladonna*. Levyns, in the recent Guide to the Flora of the Cape Peninsula, states that it grows in "bushy places on the flats and lower slopes; flowering abundantly after a fire." The blooming season is given as February to April. With me at Ithaca, New York, it blossoms in late summer and early autumn. It is a choice subject with its close umbel of shell-pink flowers, when no leaves are showing.

To Linnæus Amaryllis was a genus of nine species in the first edition of Species Plantarum in 1753, and of eleven species in the third edition, 1764. In both editions the Belladonna was ascribed to the Caribbean region, Barbados and Surinam. Philip Miller spoke of it as Mexican. There is an account of *Amaryllis Belladonna* in Botanical Magazine, 1804, in which it is stated that the plant came to England in 1712 from Portugal, but where native was yet doubtful, but "the channel through which the plant has been received makes it more than probable that it is a Brazil vegetable." A variety with pale flowers is said to have come from the Cape of Good Hope. It was long before the nativity was cleared up. Even as late as Nicholson's Illustrated Dictionary of Gardening, 1888, it was given as West Indian, perpetuating the old garden tradition.

The plant was correctly stated to be native of the Cape of Good Hope by William Herbert in his standard work on the Amaryllidaceæ in 1837; he wrote that it was naturalized in Madeira, "having been probably disseminated from gardens."

The plant we know as Belladonna lily (which is not a lily) or *Amaryllis Belladonna,* was apparently widely spread in cultivation before Linnæus wrote. The nativity of the plant was indefinitely stated or taken for granted, as with many other cultivated subjects. Linnæus cites the *Lilium rubrum* of Merian under his *Amaryllis Belladonna,* and also a plate in Albertus Seba's Thesaurus of 1734. There is a beautiful plate in Maria Sibilla Merian's delightful work, published at Amsterdam in 1726, being a dissertation on the insects, worms, lizards, caterpillars, serpents, fishes, plants, flowers, fruits and other things of Surinam (Dutch Guiana) all in wonderful colored work; with the great book open on the table I am impressed with the joy the authoress must have experienced in days before there needed to be entomologists and herpetologists and botanists and ichthyologists and all the others and when nature presented itself as a single scene of life and everything was worth recording. Well; this splendid plate No. 22 that Linnæus cites as *Lilium rubrum* does not have that designation on it or in the text, of the copy before me; it is what we now know as a Hippeastrum, and apparently *H. pu-*

niceum (or *equestre*) as is attested by Botanical Magazine, 1804, in the cited account of Belladonna lily.

This Botanical Magazine account throws an interesting sidelight on early geography notions. In speaking of Belladonna it adds: "The older Botanists call its country India, which with them may mean the East-Indies, South-America, or even some parts of Africa."

Western hemisphere plants of this relationship were separated by Herbert in a new genus, Hippeastrum; and it is to this name that the usual amaryllises of florists and of catalogues, the bulbs of which are common in the markets, are to be referred. They have been much modified by cultivation and perhaps by crossing, but most of them are of the *Hippeastrum Reginæ* class although I see *H. puniceum (equestre)* in gardens in southern parts of the United States and in the American tropics. It is not commonly known that the amaryllises of window-gardens and pots are a very different kind of plant from the real amaryllis.

In saying that the true amaryllis is South African I am following customary botanical interpretation. Herbert states that *Amaryllis Belladonna* is the type of the genus, and since his time we have regarded the genus as monotypic and have applied the name to the South African plant. By what process he arrived at that conclusion I do not know. There is nothing in the Linnæan account to singularize this species. Linnæus refers to Hortus Cliffortianus, where are references to Hermann, who died in 1695 and wrote of "Lilium americanum, puniceo flore, bella donna dictum," and to Plukenet, 1720, who had "Lilio Narcissus americanus, puniceo flore, Bella donna dictus." In both these accounts the American source is indicated as well as the reddish-purple (puniceus) nature of the flowers and the fact that the plant was called bella donna. These references, if they can be identified at all, are very likely what we now know as *Hippeastrum puniceum,* same as the plate in Merian also cited by Linnæus. This *puniceum* name comes about in this way: *Amaryllis punicea,* Lamarck (1783), which is supported by the picture in Hermann's Paradisus and in Merian, and which is *Amaryllis equestris* of Aiton (1789) and *Hippeastrum equestre* of Herbert (1821) and which Urban in 1903 brought over as *Hippeastrum puniceum* to displace the name *equestre* on the basis of priority. There is no specimen of *Amaryllis Belladonna* in the Linnean herbarium to identify his plant.

It is not my purpose in this writing to endeavor to determine the proper interpretation of Linnæus' *Amaryllis Belladonna,* but only to acquaint the reader with the kinds of problems that arise in so many of these old cases.

The word amaryllis is of course a classical name of a shepherdess or country maiden, fancifully applied to these plants. Belladonna, "beautiful lady," is an herbalist name, preserved by Linnæus in the binomial *Amaryllis Belladonna.* Just why it was given to this plant by Hermann, Plukenet and others I do not know but presumably in compliment to the handsome bloom.

There is another Belladonna, a very different plant and of which there is record of the name. This is *Atropa Belladonna,* of the Nightshade family, also a Linnæan species. The plant is powerfully poisonous, and well-known drugs are prepared from it. Symptoms of belladonna poisoning are stated by the United

States Dispensatory to be the same as those of atropine poisoning of which it is said the most striking "is the peculiar delirium. In the earlier stages this manifests itself simply by profuse and somewhat incoherent talkativeness, but later there is complete obfuscation often with hallucinations, sometimes more or less maniacal in character." The maniacal and lethal character are suggested in the identificational references quoted by Linnæus: Bauhin, *Solanum maniacum multis;* Clusius *Solanum lethale.* The name belladonna comes to the plant from the use of the red sap by women of Italy as a cosmetic.

It is now a favorite notion in some quarters that "systematic botany" has reached its end. This reveals delightful innocence of natural history. It is perhaps born in part of the present devotion to indoor laboratory work, furthered by the remarkable advances in appliances and technique and the stimulating discoveries. All that work is beyond praise. Yet the fields and hills are just outside, teeming with life, much of which is yet little known and all of which requires study in a new way. It is fashionable to deprecate the making of new species: what, then, shall we do with them,—let them go undescribed and unnamed?

Systematic botany and zoölogy, like all other subjects, have quickly responded to the evolution point of view, and life histories acquire new significance. Many of our questions are to be answered, in the end, in the field.

The kind of animals and plants must be distinguished. This is a pre-requisite to the most significant study in morphology, physiology, ecology, heredity, distribution. In fact, much of the biological work is inexpressible except in terms of species and varieties; and these categories are not cabinet conceptions. Systematics is today one of the freshest and most inspiring of the biological groups. Every advance in physiology and genetics makes it more interesting and important. The monographing of groups in the contemporary spirit is one of the new promising lines of research. More than that, natural history is not outlived, although the outlook of the workers may have changed as it has in geology and psychology. Museum specimens are not mere dead property; they are the records and symbols of living things far and wide.

The earth still has its charm. Plants will be sought and admired, scrutinized and named, to the end of man's time.

First requisite in natural history is to recognize the forms of life. This recognition must be afield, where the organisms live and multiply. Records must be kept. Forms of life are yet imperfectly known. The great laboratory is still out-of-doors; we have no reason to expect it will ever be otherwise.

The chain of life comprises not alone organisms now living on the earth. It connects with the dead and fossil past. From first to last, beginning unto end, the chain is a continuous series, a connected problem. Throughout the vast extent, study of the kinds, taxonomy, nomenclature, systematic speculation, constitute a field of ever-increasing vitality and importance.

Now is this writing finished; there follow only the lists of names, together with the necessary explanations. I hope it has left the reader with some feeling of respect for the names by which plants are known, and some comprehension of the breadth of the subject and the problems that arise.

The writing falls far short of its purpose if it does not also suggest great ranges of interest and need of investigation in the field of cultivated plants. With all the priceless researches, there still remains an undeveloped domain of biological inquiry into the origins, identities, development and essential characteristics of the plants by which mankind has sustained and comforted itself. The origins lead much farther back than recorded history, into archeology and even geology, and this field is yet little explored.

In the meantime we need records. We have means and agencies for continuous record in any number of other fields, from postage stamps to Indian lore and relics, of birds and mammals and insects and fishes and plants of the field, of books independently of their values, of every kind of device new, old and discarded. Every artifact from excavations is saved. Yet we have no recognized and sufficient archives of cultivated plants. The plants themselves, competently preserved, together with memoranda attached to them and the special experience of them accumulating generation by generation, comprise the major and indispensable chronicle. What will other generations think of us, when they find it necessary to try to pick out origins and courses of amelioration, and positive evidences of introductions and novelties, from these our headlong days?

VI

THE NAMES AND THE WORDS

BOTANICAL nomenclature is Latin. Thereby may it be understandable to all peoples in all languages. This nomenclature is a combination of nouns and adjectives. Verbs and other forms of speech are employed in Latin descriptions, but not in the names.

First name in the binomial is a substantive (noun), nominative case and singular number; second name is usually an adjective, modifying the substantive. Tree is a substantive; tall, low, young, old, beautiful, are adjectives suggesting the kind or quality of a particular tree.

All words are beautiful when properly used and correctly pronounced and relieved of the vulgarisms of slang. So the binomials of plants and animals are beautiful if clearly enunciated and decently pronounced. They constitute a luminous part of the language of horticulture, botany, and natural history.

For the most part, these binomials are not difficult in speech. Of course practice is required to speak any vocabulary well, whether of art, engineering, architecture, music, medicine, education or law; accurate clear language is the mark of sensitiveness and intelligence.

Use of these binomials is good training in precision of speech. They are dignified and maintain themselves above the mumbling and mixture of daily language. Thus a bookful of special words, such as Standardized Plant Names, has something of the appeal to certain persons that a score of music has to others.

First must one comprehend the genus, as Acer, the maples, is a genus of many species, Rosa, Chrysanthemum, Magnolia, Prunus, Berberis, genera also of many species. A plant very distinct in essential characters from all other plants may constitute a genus by itself, as ginkgo, heather, amaryllis, coconut; monotypic genera are named binomially same as the others, the cited cases being *Ginkgo biloba, Calluna vulgaris, Amaryllis Belladonna, Cocos nucifera*, other binomials in the last two genera not belonging there.

Latin is an inflected language, by which it is meant that a word changes form to express relations or different genders. Thus, nouns ending in *us* when the subject of a sentence end in *um* when the object, although this grammatical change does not interest us in nomenclature. More to the point in our problem is the fact that nouns have gender, and all nouns are masculine, feminine or neuter; gender in this case is not necessarily an attribution of sex but is rather a form of language. Adjectives do not have gender, but correspond with their nouns in this respect. Thus, *Ceanothus americanus* is a masculine name, *Cimicifuga americana* feminine, *Narthecium americanum* neuter.

Agreement for gender in the two members of the binomial does not always result in endings identical in both genus and species. Thus the adjective meaning white is *albus, -a -um,* masculine, feminine and neuter respectively, while black is *niger, -a, -um: Helleborus niger, Brassica nigra, Solanum nigrum; ruber* (red) is a similar case.

Comparable with *niger* and *ruber* in ending are certain *-fer* and *-ger* words meaning "bearing": *umbellifer*, masculine, *umbellifera*, feminine, *umbelliferum*, neuter; *setiger, setigera, setigerum*. It would not be allowable to terminate such words in *us* for masculine, although cases occur. Apparently more of this class of adjectives occur with feminine nouns than with others and they are entered in feminine form in the following lists although otherwise masculine terminations are given.

The substantives (generic names) are classical Latin names, often originally derived from Greek, or words compounded of Latin or Greek, or of other origin and more or less Latinized in form: the point is that the names are sufficiently adopted into Latin to be declined and readily used in technical diagnoses. Often they commemorate persons, as Linnæa, Bauhinia, Parkinsonia, Dodonæa, Clusia, Besleria, Tournefortia, Milleria; frequently they are classical words with a general or even indefinite meaning that have been applied in botany to a particular group of plants or even to a different group from that originally intended; these new applications in no way invalidate them as generic names. Celastrus was employed in Greek for some kind of evergreen, Ilex is Latin for a kind of oak, as also Æsculus, Hypericum is of obscure classical signification, Lycium was applied to a Rhamnus. Probably more than one half the generic names one commonly meets are of classical origin, meaning Greek and Latin.

If an author is not obliged to retain the original

meaning of the word he adopts for a genus, neither is he obligated to accept its exact spelling. It is legitimate for Linnæus to write Buddleja, named in memory of Adam Buddle. Neither is an author obliged to spell a generic appellation in the same way as does the person for whom it was named. Thus Kennedia was named after Lewis Kennedy, British nurseryman, but Ventenat who founded the genus preferred to write it in regular Latin form; Stewartia was named for John Stuart, Earl of Bute; Stillingia memorializes Dr. Stillingfleet. Botanists chose to modify the personal name Euphorbus to Euphorbia. Generic names derived from persons are not primarily commemorative. Neither the horticulturist nor the botanist need give much attention to the literal significance of the genus-names except as a matter of interest or information : a name is a name for all that.

So is a specific adjective a name for all that; but the literal meaning of the word becomes part of the background in the language of botany. It aids considerably to know that *Betula lutea* means yellow birch, *B. lenta* pliable or flexible birch, *B. pumila* dwarf birch, *B. populifolia* poplar-leaved birch, *B. papyrifera* paper-bearing birch, although it may not serve useful purpose to use translations as English names; nor is there any guaranty that the name is really applicable in a given case, as witness *Duranta repens* (repent or creeping) for an upright big shrub, with only some of the branches more or less lying on the ground.

Although the orthography is not to be changed, yet the termination of adjectives must naturally follow the gender of the generic noun. Thus a dwarf sunflower is called *Helianthus pumilus* (masculine), a dwarf birch *Betula pumila* (feminine), a dwarf daisy *Chrysanthemum pumilum* (neuter).

In the long lists that follow, the generic names (List I) are merely pronounced; the specific names (List II) are pronounced and also the meaning suggested. One soon learns the significance of the species-names by frequently consulting such lists, if one has feeling for words. In some cases, however, care must be taken to distinguish. Thus *Dianthus macranthus* is long-flowered or large-flowered pink, but *Acacia macracantha* is long- or large-spined acacia, one termination being from Greek *anthos*, flower, and the other *acanthos*, spine or thorn.

The *macro-* words need explanation, as *macrocarpus* referring to fruit, *macrodontus* to teeth, *macromeris* to parts, *macrospermus* to seeds, *macrostachyus* to spikes. The Greek *macros* means long, yet in botanical practice the combinations commonly signify large, big, great, in distinction from *micros*, small. Thus *Aster macrophylla* is interpreted Bigleaf aster, *Philadelphus microphyllus* the Littleleaf mock-orange. This practice follows good accepted English usage, as macrophone and microphone, macrocosm and microcosm, macroscope and microscope.

Another contrast is *Salix cordifolia*, heart-leaved willow, and *Æthionema coridifolium*, coris-leaved, Coris being a genus in the Primrose family. The latter binomial may seem to be a case of gender disagreement between genus and species, but Æthionema is a Greek neuter, as are Aglaonema, Odontonema. Masculine Greek endings *os* become *us* when Latinized, but the original author has choice as to which form he uses in noun or adjective. Thus Siebold & Zuccarini founded the genus Rhodotypos, which many succeeding authors write Rhodotypus. There are many comparable cases. Similarly, an author may choose a Greek neuter termination rather than to Latinize it *um:* example is *Asplenium platyneuron.*

In generic names one must also be careful not to confuse those of very similar spelling. The Rules provide that such names are not duplicates if they differ by as much as one letter. We have different genera with such similar names as Discocactus and Disocactus, Jaegeria and Jagera, Nolana and Nolina, Lomatia and Lomatium, Butea and Butia, Ceropteris (Pityrogramma) and Ceratopteris and Cystopteris, Garberia and Gerberia, Morinda and Moringa, Syringa and Seringia, Ligustrum and Ligusticum, Anemopsis and Anemonopsis, Latania and Lantana.

The consultant may not be interested in these reasons and differences but they emphasize the fact that one must be careful to follow the spelling in authoritative lists and books.

Adjectival names may be made from the titles of countries or regions: *Anemone virginiana,* Virginian anemony; *Iris virginica,* Virginican iris; *Saxifraga virginiensis,* pertaining to (citizen of) Virginia. These spellings are to be retained as they were first published: the different adjectival forms are not interchangeable even though their significance may be the same. Frequently the application or origin of geographical names is not at once apparent: *Aconitum noveboracense, Vernonia noveboracensis,* citizens of New York (*Eboracum,* Roman name of York, England, *novum,* new).

Sometimes these geographical names go far astray. We have noted the case of Portugal cypress, implied in the name *Cupressus lusitanica,* which is native in

Mexico (page 4). The common big milkweed of eastern fields is *Asclepias syriaca,* so named by Linnæus from old designations of it, although he himself knew that it is Virginian: it was *Apocynum majus syriacum rectum* of Cornut (Cornutus) who published on the plants of Canada in 1635, and *Apocynum syriacum* of Clusius. Because all species of Asclepias are native in the New World, Decaisne in 1844 renamed this plant *Asclepias Cornuti* and so it was known for a long time, but we must go back to the old name and be content that it records an early misapprehension. Point is that a name is a name independently of its literal meaning; and in the big catalogue that follows (List II) the meanings are given only as information.

The same geographical name may be differently spelled or one country may have two appellations: *Rosa sinica, Rosa cathayensis,* Chinese roses, quite distinct species (the former chanced to have been earlier named *R. lævigata); Juniperus chinensis,* Chinese juniper or cedar; *Citrus sinensis,* Chinese orange. Any of these different forms could hold, without conflict, even if made in the same genus, although unfortunate, as also *Ligustrum japonicum,* Japanese privet, and *Chrysanthemum nipponicum,* Japanese (Nipponese) chrysanthemum.

Certain adjectives are formed by the Greek termination *-oides, -oideus, -ides, -odes,* meaning like or resembling: *Epiphyllum phyllanthoides,* a phyllanthus-like epiphyllum; *Canna orchiodes,* orchid-like canna; *Populus deltoidea,* leaves delta-like (triangular).

Not all specific names are common adjectives. Frequently they are proper nouns in the genitive, equivalent to the English possessive. *Phlox Drum-*

mondii is the phlox of Drummond or Drummond's phlox. The genitive is formed in several ways, depending on the declension into which the substantive falls. If a personal name is assumed to terminate in *us* when Latinized, as is usual, thus making it second declension, the genitive would be in *i*. Thus do we have Linnæus and Linnæi, Clusius and Clusii, Dodonæus and Dodonæi. There is variable practice as to whether the genitive shall be formed by *i* or by *ii*. On this point the International Rules of Nomenclature recommend that when the personal name ends in a vowel, the letter *i* is added; when it ends in a consonant (except in *r*) the letters *ii* are added; this recommendation is not directly retroactive.

Names of women, ordinarily assumed to be of the first declension and ending in *a*, take *æ* for the genitive; *Rosa Banksiæ*, Lady Banks' rose.

Certain genitives, as in the third declension, are made in *is: Rosa Hugonis*, Hugo's rose; *Solidago ohionis*, goldenrod of Ohio.

Genitives are sometimes made in the plural: *Colocasia antiquorum*, colocasia of the ancients (antiqui); *Grimaldia Baileyorum*, of the Baileys (father and daughter).

Frequently, genitives (possessives) are formed from compound geographical names: *Aster novæ-angliæ*, New England aster; *A. novi-belgii*, New Belgian (New York) aster; *Lechea novæ-cæsareæ*, New Jersey pinwort (Cæsarea, Roman name of the Channel Islands from which the word Jersey is derived); *Rubus pergratus* var. *novæ-terræ*, Newfoundland blackberry.

If a botanist desires to name a new species in compliment to a person, two regular ways are available: to make a genitive of the noun, as *Smithii* or *Smithiæ,* whether the person is masculine or feminine; to put the name in adjectival form, as *Smithianus, Smithiana, Smithianum,* whether the genus is masculine, feminine, or neuter.

Certain specific names lie outside the regular rules. These are nouns in apposition, and cannot be altered to agree in gender. Usually they are historic substantives that have come down in the literature of the subject: *Rumex Patientia,* the old herb-patient, a medicinal plant; *Chenopodium Bonus-Henricus,* the Good King Henry of the herbalists; *Nicotiana Tabacum,* preserving the aboriginal name of tobacco; *Solanum PseudoCapsicum; Thymus Serpyllum; Aconitum Anthora;* or an old generic name as *Persica* is for the peach and we write *Prunus Persica,* but the same word is merely a geographical adjective in other cases, as in *Syringa persica* (page 60). Such substantive names are preferably retained with a capital initial, to indicate that they are not adjectives and to preserve their importance.

Some writers prefer to use no capitals in specific names, not even in those commemorating persons, writing *Salvia greggii, Pyrus halliana, Pinus jeffreyi.* This is in the interest of uniformity; but uniformity, which is the fetish of standardization, has no supernatural merit. It is much more desirable to maintain dignity and emphasis than to insist on the flatness of regularity. Suggestion of much precious history is lost when the identifying capitals are deposed.

Formerly, specific names of countries were written with capital initials, as Canadensis, Japonica,

Africana, but this custom is not now universal. Geographic epithets are scarcely proper names in these cases, but have come to represent general regions of nativity. *Rubus canadensis* is not exclusively Canadian; it grows as far south as Georgia; the name indicates where Kalm first discovered it. In early days, when many of our plants were named, Virginia was much more than the present state of Virginia; Brazil was a region or direction in the western hemisphere. Because a *Potentilla* was named *pennsylvanica* does not cause the collector to be surprised to find it in New Hampshire, Ontario or Oregon; I have taken it in China, for it is put down as native Caucasus to Japan; this world-wide species happens to have been named and described from Pennsylvania in 1767 by Linnæus. Regional adjectives hardly merit great distinction; but personal proper names and rich old substantives in apposition may well be allowed the justice of a capital initial.

Practice in capitalizing species-names is not mandatory in rules of nomenclature. International Rules recommend that "specific names begin with a small letter except those which are taken from names of persons (substantives or adjectives) or those which are taken from generic names (substantives or adjectives)." American Code provides that "if capital letters are to be used for specific names they should be employed only for substantives and for adjectives derived from personal names."

PRONUNCIATION

In the lists that follow no effort is made to indicate complete pronunciation. That attempt would require diacritical marks or a phonetic alphabet.

Only two purposes are in mind: the accentuation, by which the syllable of primary accent or stress is indicated; quantity of the accented vowels, as to whether "long" or "short."

There is no standard agreement on rules for the pronunciation of botanical binomials. Even in the best practice, there may be variations in pronunciation of a given word; this is unavoidable, and no more to be regretted than similar variations in pronouncing many English words. The particular sound to be given the vowels (within the categories "long" or "short") rests with the individual. Many persons pronounce generic and specific names simply as if the words were English, but for the most part the accent, at least, follows usage in Latin.

Good examples of different pronunciations in Latin (derived from Greek) and English are the *-oides* terminations (which we have met on a preceding page). In English, *oi* under similar circumstances is a diphthong, as in rhomboid, pronounced like *oy* in toy; in Latin or Latinized nomials *oi* is not a diphthong but two separate vowels.

It may be said there are two ways of pronouncing Latin. One is the so-called Roman method followed by Latinists, that represents what is considered to be the pronunciation of classical times; the other is an adaptation of pronunciation more or less to the speech of people now using it. It is only the latter with which we are concerned in this discussion.

In the United States as well as in England, the vowels are usually pronounced with English sounds. This means that the long English *i* and *e* (which are singularities among languages) may be used. In the United States I cannot be corrected if I say Lupinus with a long English *i;* in France I should say Lupeenus. Collecting far away in South America with a delightful companion who spoke a Latin language, I came upon plants of Sida and pronounced the word with long English *i,* whereupon my comrade noted my lack of understanding and corrected me to Seeda; it was not necessary to remind him that my native speech is English. So, also, whether one pronounces *americana* as if it were spelled *americay-na* or *americah-na* depends on choice, or perhaps whether one is from Boston or Kansas; my own habit is the former, although subject to suggestion. Either way the vowel may be considered as "long."

Terminal syllables of the natural families are commonly pronounced in the United States as if they were English: *Rosaceæ,*—àce-ee, with long or open sounds for *e.*

The foregoing remarks have reference particularly to the pronunciation of letters, not so much to accent of words. Accent or stress follows rules of Latin; and the syllables are as many as the vowels. Words of two syllables are stressed on the first syllable: *à-cris;* of three syllables on next-to-the-last syllable (penult) if it is long: *dumò-sa;* if it is short, accent may be on the preceding syllable or antepenult, but never on a syllable before the third from the end.

Inasmuch as many names, particularly of genera, are derived from non-Latin sources and may be only imperfectly Latinized, it is impossible to follow rules steadfastly. How the accents fall in particular words is indicated in the lists that follow, although there may be disagreement in some cases.

Let it be repeated that the pronunciations here suggested follow practice in the United States. To indicate the quantity of the vowel (whether long or short) accent-marks are employed, to the left for long open articulation and to the right for short close sounds. This is now an American custom, although of English origin. Thus Asa Gray, in his first Manual of the Botany of the Northern United States, 1848, writes: "To aid in their pronunciation, I have not only marked the accented syllable, but have followed Loudon's mode of indicating what is called the long sound of the vowel by the grave (`), and the short sound by the acute (´) accent-mark." In the preface to his Hortus Britannicus, first published in 1830, J. C. Loudon explains his method of pronunciation. The current (Seventh) edition of Gray's Manual, 1913, by his successors, maintains this use of the accent-marks for vowel quantity and also for syllable stress, although not adopted in the Synoptical Flora. American botanical practice is not wholly uniform, but in the present lists the custom long established by Gray and his followers is adopted.

It is difficult to represent pronunciation by means of such simple marks and there are many exceptions, particularly in words derived from personal and geographic names and in those not known in classical Latin.

The specific or trivial names in List II are an extension of a similar compilation on pages 148 to 159 of the first volume of Standard Cyclopedia of Horticulture and repeated in part on pages 21–36 of Manual of Cultivated Plants; the List has therefore had the test of previous review, although nothing like perfection can be expected of it.

Variable practice obtains in the pronunciation of names made from those of persons, particularly when the patronymic is in two syllables. Probably the Latin preference is to accent on the penultimate, but frequently the words are spoken as the persons pronounced their own names. This applies both to genitives as specific names and to substantives as generic names. Thus one may say Tór-reyi, Tór-reya rather than Torrèy-i and Torrèy-a. Similar cases are Búck-leyi or Bucklèy-i, Búck-leya or Bucklèy-a, Jà-mesii or Jamè-sii, Jà-mesia or Jamè-sia. It is the intention to omit most names of this character from the Lists. English-speaking horticulturists, as far as I have noted them, say Cátt-leya rather than Cattlèy-a.

Finally let it be said that the following lists are compiled primarily for the horticultural fraternity. They are not final or at least not infallible and are subject to revision as needed.

LIST I

Generic names likely to be met in horticultural literature, with indication of accent and vowel quantity, and ready reference in spelling.

Grave accent (`` ` ``), means long vowel;
acute accent (`` ' ``), short or similar
vowel sounds, or at least not long

Abè-lia
À-bies
Abò-bra
Abrò-ma
Abrò-nia
Abrophýl-lum
Ἀ brus
Abù-tilon
Acà-cia
Acæ-na
Acalỳ-pha
Acám-pe
Acanthocè-reus
Acantholì-mon
Acanthóp-anax
Acanthophœ̀-nix
Acanthophýl-lum
Acanthorhì-za
Acán-thus
À-cer
Acerán-thus
Achillè-a
Achím-enes
À-chlys
Ách-ras
Acidanthè-ra
Acinè-ta
Ackà-ma
Accœlorrà-phe

Acokanthè-ra
Aconì-tum
Ác-orus
Acrocò-mia
Acroných-ia
Actæ̀-a
Actiníd-ia
Actinophlœ̀-us
Actinós-trobus
À-da
Adansò-nia
Adelocalým-na
Adelocà-ryum
Adenanthè-ra
Adenocár-pus
Adenóph-ora
Adenós-toma
Adhát-oda
Adián-tum
Adlù-mia
Adoníd-ia
Adò-nis
Adóx-a
Æchmè-a
Æ-gle
Æglóp-sis
Ægopò-dium
Æò-nium
Ærán-gis

Aerì-des
Ǽr-va
Ǽs-culus
Æthionè-ma
Agapán-thus
Agás-tache
Ág-athis
Agathós-ma
Agà-ve
Agdés-tis
Agér-atum
Aglaonè-ma
Agò-nis
Agrimò-nia
Agrostém-ma
Agrós-tis
Aichrỳ-son
Ailán-thus
Aíph-anes
Aì-ra
Ajù-ga
Akè-bia
Albíz-zia
Alchemíl-la
Aléc-tryon
Ál-etris
Aleurì-tes
Alís-ma
Allagóp-tera

119

Allamán-da
Alliò-nia
Ál-lium
Allóph-yton
Allopléc-tus
Ál-nus
Alocà-sia
Ál-oë
Alonsò-a
Alopecù-rus
Alphitò-nia
Alpín-ia
Alseuós-mia
Alsóph-ila
Alstò-nia
Alstrœmè-ria
Alternanthè-ra
Althǽ-a
Alýs-sum
Alýx-ia
Amarà-cus
Amarán-thus
Amárc-rinum
Amarýl-lis
Amasò-nia
Amberbò-a
Amelán-chier
Amél-lus
Amhér-stia
Amián-thium
Amíc-ia
Ammò-bium
Ammóch-aris
Ammóph-ila
Amò-mum
Amór-pha
Amorphophál-lus
Ampelodés-ma
Ampelóp-sis
Amphíc-ome
Amsò-nia
Anacámp-seros
Anacár-dium

Anacỳ-clus
Anagál-lis
Anán-as
Anáph-alis
Anastát-ica
Anathè-rum
Anchù-sa
Andì-ra
Andróm-eda
Andropò-gon
Andrós-ace
Androstè-phium
Anemò-ne
Anemonél-la
Anemonóp-sis
Anemopǽg-ma
Anemóp-sis
Anè-thum
Angél-ica
Angelò-nia
Angióp-teris
Angóph-ora
Angrǽ-cum
Angulò-a
Anigozán-thos
Anisacán-thus
Anisót-ome
Annò-na
Anò-da
Anóp-teris
Anò-ta
Ansél-lia
Antennà-ria
Án-themis
Anthér-icum
Antholỳ-za
Anthoxán-thum
Anthrís-cus
Anthù-rium
Anthýl-lis
Antià-ris
Antidés-ma
Antíg-onon

Antirrhì-num
Aphanós-tephus
Aphelán-dra
À-pios
À-pium
Apléc-trum
Apóc-ynum
Aponogè-ton
Aporocác-tus
Aptè-nia
Aquilè-gia
Ár-abis
Ár-achis
Arách-nis
Arà-lia
Araucà-ria
Araù-jia
Ár-butus
Archontophœ-nix
Árc-tium
Arctostáph-ylos
Arctò-tis
Arctò-us
Ardís-ia
Arè-ca
Arecás-trum
Aregè-lia
Arenà-ria
Arén-ga
Arethù-sa
Argà-nia
Argemò-ne
Argyrè-ia
Aridà-ria
Arikuryrò-ba
Ariocár-pus
Arisǽ-ma
Arís-tea
Aristolò-chia
Aristotè-lia
Armorà-cia
Arnè-bia
Ár-nica

Aròn-nia
Arpophýl-lum
Arracà-cia
Arrhenathè-rum
Artáb-otrys
Artemís-ia
Arthropò-dium
Artocár-pus
À-rum
Arún-cus
Arundinà-ria
Arún-do
Ás-arum
Ascár-ina
Asclè-pias
Asclepiodò-ra
Ascocén-trum
Ascotaín-ia
Ás-cyrum
Asím-ina
Aspár-agus
Aspér-ula
Asphodelì-ne
Asphód-elus
Aspidís-tra
Aspidospér-ma
Asplè-nium
Ás-pris
Astartè-a
Astè-lia
Ás-ter
Astíl-be
Astrág-alus
Astrán-tia
Astrocà-ryum
Astróph-ytum
Asystà-sia
Atalán-tia
Athamán-ta
Athrotáx-is
Athýr-ium
Atrapháx-is
Át-riplex

Át-ropa
Attalè-a
Aubriè-ta
Aucù-ba
Audibér-tia
Audouín-ia
Aureolà-ria
Avè-na
Averrhò-a
Axón-opus
Azà-ra
Azól-la

Babià-na
Bác-charis
Bác-tris
Baè-ria
Baillò-nia
Balà-ka
Balaù-stion
Ballò-ta
Balsamocít-rus
Balsamorhì-za
Bambù-sa
Bánk-sia
Báph-ia
Baptís-ia
Barbarè-a
Bárk-lya
Barlè-ria
Barós-ma
Barringtò-nia
Basél-la
Bauè-ra
Bauhín-ia
Beaucár-nea
Beaufór-tia
Beaumón-tia
Befà-ria
Begò-nia
Belamcán-da
Belepér-one
Bél-lis

Bél-lium
Benincà-sa
Bén-zoin
Berberidóp-sis
Bér-beris
Berchè-mia
Bergè-nia
Bergerán-thus
Bergerocác-tus
Berlandiè-ra
Berterò-a
Berthollè-tia
Bertolò-nia
Bè-ta
Bét-ula
Bì-dens
Bifrenà-ria
Bignò-nia
Billardiè-ra
Billbér-gia
Bischóf-ia
Biscutél-la
Bismár-ckia
Bíx-a
Blanfór-dia
Bléch-num
Bletíl-la
Blì-ghia
Bloomè-ria
Blumenbách-ia
Boccò-nia
Bœhmè-ria
Boisduvà-lia
Boltò-nia
Bolusán-thus
Bomà-rea
Bóm-bax
Bón-tia
Borà-go
Borás-sus
Borò-nia
Bortých-ium
Bò-sea

121

Bossiæ-a
Boussingaúl-tia
Bouvár-dia
Bowkè-ria
Boykín-ia
Brachých-iton
Brachýc-ome
Brachyglót-tis
Brachypò-dium
Brachysè-ma
Brà-hea
Brasè-nia
Brassaocattlæ-lia
Brassáv-ola
Brás-sia
Brás-sica
Brassocátt-leya
Brassolæ-lia
Brevoór-tia
Brèy-nia
Brickél-lia
Brittonás-trum
Brì-za
Brodiæ-a
Bromè-lia
Brò-mus
Brós-imum
Broughtò-nia
Broussonè-tia
Browál-lia
Brów-nea
Bruckenthà-lia
Brunnè-ra
Brunsfél-sia
Brunsvíg-ia
Bryò-nia
Bryonóp-sis
Bryophýl-lum
Buckleỳ-a
Buddlè-ja
Buginvíl-læa
Bulbì-ne
Bulbinél-la

Bulbocò-dium
Bulbophýl-lum
Bumè-lia
Buphthál-mum
Bupleù-rum
Bursà-ria
Bù-tia
Bù-tomus
Búx-us
Byrnè-sia

Cabóm-ba
Cæsalpì-nia
Cailliè-a
Caióph-ora
Cajà-nus
Calacì-num
Calà-dium
Cál-amus
Calandrì-nia
Calán-the
Calathè-a
Calceolà-ria
Calén-dula
Calím-eris
Cál-la
Callián-dra
Callicár-pa
Callíc-oma
Callír-hoë
Callistè-mon
Callís-tephus
Callì-tris
Callù-na
Calocéph-alus
Calochór-tus
Calodén-drum
Calonýc-tion
Calóph-aca
Calophýl-lum
Calopò-gon
Calothám-nus
Calpúr-nia

Cál-tha
Calycán-thus
Calycót-ome
Calýp-so
Calỳ-trix
Camarò-tis
Camá-sia
Camél-lia
Camoén-sia
Campán-ula
Camphorós-ma
Campsíd-ium
Cámp-sis
Camptosò-rus
Camptothè-ca
Campylót-ropis
Canán-ga
Canarì-na
Canavà-lia
Candól-lea
Canél-la
Canís-trum
Cán-na
Cán-nabis
Cán-tua
Cáp-paris
Cáp-sicum
Caragà-na
Cardám-ine
Cardián-dra
Cardiospér-mum
Cár-duus
Cà-rex
Cà-rica
Carís-sa
Carlì-na
Carludovì-ca
Carmichæ-lia
Carnè-giea
Carpán-thea
Carpentè-ria
Carpì-nus
Carpobrò-tus

Carpód-etus
Carriè-rea
Cár-thamus
Cà-rum
Cà-rya
Caryóp-teris
Caryò-ta
Casimír-oa
Cás-sia
Cassín-ia
Castà-nea
Castanóp-sis
Castanospér-mum
Castíl-la
Castilè-ja
Casuarì-na
Catál-pa
Catanán-che
Catasè-tum
Catesbǽ-a
Cà-tha
Cathcár-tia
Cátt-leya
Caulophýl-lum
Ceanò-thus
Cecrò-pia
Céd-rela, Cedrè-la
Cedronél-la
Cè-drus, Céd-rus
Ceì-ba
Celás-trus
Celmís-ia
Celò-sia
Cél-sia
Cél-tis
Centaurè-a
Centaù-rium
Centhrán-thus
Centradè-nia
Centropò-gon
Centrosè-ma
Cephäë-lis
Cephalán-thus

Cephalà-ria
Cephalocè-reus
Cephalostà-chyum
Cephalotáx-us
Cerás-tium
Ceratò-nia
Ceratopét-alum
Ceratophýl-lum
Ceratóp-teris
Ceratostíg-ma
Ceratozà-mia
Cercidiphýl-lum
Cercíd-ium
Cér-cis
Cercocár-pus
Cè-reus
Cerín-the
Ceropè-gia
Ceróx-ylon
Cés-trum
Chænomè-les
Chænós-toma
Chærophýl-lum
Chamœcè-reus
Chamæcýp-aris
Chamædáph-ne
Chamædò-rea
Chamælaù-cium
Chamælír-ium
Chamǽ-rops
Chambeyrò-nia
Chár-ieis
Cheilán-thes
Cheirán-thus
Chelidò-nium
Chelò-ne
Chenopò-dium
Chilóp-sis
Chimáph-ila
Chiocóc-ca
Chióg-enes
Chionán-thus
Chionodóx-a

Chionóph-ila
Chirò-nia
Chlò-ris
Chlorocò-don
Chloróg-alum
Chloróph-ora
Chloróph-ytum
Choís-ya
Chorís-ia
Choríz-ema, Cho-
 rizè-ma
Chrysalidocár-pus
Chrysán-themum
Chrysobál-anus
Chrysóg-onum
Chrysóp-sis
Chrysosplè-nium
Chrysothám-nus
Chusquè-a
Chỳ-sis
Cibò-tium
Cì-cer
Cichò-rium
Cicù-ta
Cimicíf-uga
Cinchò-na
Cinnamò-mum
Cipù-ra
Circǽ-a
Cír-sium
Cís-sus
Cís-tus
Citharéx-ylum
Citróp-sis
Citrúl-lus
Cít-rus
Cladán-thus
Cladrás-tis
Clár-kia
Clausè-na
Clavì-ja
Claytò-nia
Cleistocác-tus

Clém-atis
Cleò-me
Clerodén-drum
Cléth-ra, Clè-thra
Clián-thus
Cliftò-nia
Clintò-nia
Clitò-ria
Clì-via
Clytós-toma
Cneoríd-ium
Cneò-rum
Cnì-cus
Cobǽ-a
Coccín-ia
Coccocýp-selum
Coccól-obis
Coccothrì-nax
Cóc-culus
Cochemiè-a
Cochleà-ria
Cochlospér-mum
Cò-cos
Codiǽ-um
Codonóp-sis
Cǿ-lia
Cœlóg-yne
Coffè-a
Cò-ix
Cò-la
Cól-chicum
Coleonè-ma
Cò-leus
Collè-tia
Collín-sia
Collinsò-nia
Collò-mia
Colocà-sia
Colpothrì-nax
Colquhoù-nia
Colúm-nea
Colù-tea
Colvíl-lea

Combrè-tum
Comespér-ma
Commelì-na
Comptò-nia
Conán-dron
Condà-lia
Cón-gea
Conicò-sia
Coniográm-me
Conì-um
Conóph-ytum
Convallà-ria
Convól-vulus
Coopè-ria
Copaíf-era
Coperníc-ia
Coprós-ma
Cóp-tis
Cór-chorus
Cór-dia
Córd-ula
Cordylì-ne
Corè-ma
Coreóp-sis
Corethróg-yne
Corián· lrum
Corià-ria
Cór-nus
Corò-kia
Coroníl-la
Corón-opus
Corò-zo
Corrè-a
Cortadè-ria
Cortù-sa
Corýd-alis
Corylóp-sis
Cór-ylus
Corynocár-pus
Corý-pha
Coryphán-tha
Corytholò-ma
Cós-mos

Cós-tus
Cót-inus
Cotoneás-ter
Cót-ula
Cotylè-don
Coutà-rea
Cowà-nia
Crám-be
Craspè-dia
Crás-sula
Cratǽ-gus
Crè-pis
Crescén-tia
Crinodén-dron
Crinodón-na
Crì-num
Cristà-ria
Críth-mum
Crocós-mia
Crò-cus
Crossán-dra
Crotalà-ria
Crucianél-la
Crupì-na
Cryóph-ytum
Cryptán-tha
Cryptán-thus
Cryptocà-rya
Cryptográm-ma
Cryptól-epis
Cryptomè-ria
Cryptostè-gia
Cryptostém-ma
Ctenán-the
Cù-cumis
Cucúr-bita
Cù-minum
Cunì-la
Cunninghám-ia
Cupà-nia
Cù-phea
Cuprés-sus
Curcù-ligo

124

Cúr-cuma	Darlingtò-nia	Dictyospér-ma
Cyanò-tis	Darwín-ia	Dieffenbách-ia
Cýáth-ea	Dasylír-ion	Dierà-ma
Cyathò-des	Datís-ca	Diervíl-la
Cỳ-cas	Datù-ra	Digità-lis
Cýc-lamen	Daubentò-nia	Dillè-nia
Cyclanthè-ra	Daù-cus	Dillwýn-ia
Cyclán-thus	Davál-lia	Dimorphothè-ca
Cyclóph-orus	Davíd-ia	Dinè-ma
Cycnò-ches	Debregeà-sia	Dioclè-a
Cydís-ta	Decaì-snea	Dì-on
Cydò-nia	Deckè-nia	Dionæ-a
Cymbalà-ria	Déc-odon	Dioscorè-a
Cymbíd-ium	Decumà ria	Diós-ma
Cymbopò-gon	Deeríng-ia	Diospỳ-ros
Cynán-chum	Delò-nix	Diò-tis
Cýn-ara	Delospér-ma	Dipél-ta
Cýn-odon	Delós-toma	Diphyllè-ia
Cynoglós-sum	Delphín-ium	Dipladè-nia
Cynosù-rus	Demazè-ria	Diplà-zium
Cypél-la	Dendrò-bium	Diploglót-tis
Cypè-rus	Dendrocál-amus	Diplotáx-is
Cyphomán-dra	Dendrochì-lum	Díp-sacus
Cypripè-dium	Dendromè-con	Dipterò-nia
Cyríl-la	Dennstǽd-tia	Dír-ca
Cyrtò-mium	Dentà-ria	Dì-sa
Cyrtopò-dium	Dér-ris	Discár-ia
Cyrtós-tachys	Desfontaì-nea	Discocác-tus
Cystóp-teris	Desmán-thus	Disocác-tus
Cýt-isus	Desmò-dium	Disphỳ-ma
	Desmón-cus	Dís-porum
Daboè-cia	Detà-rium	Dís-tictis
Dacrýd-ium	Deù-tzia	Distỳ-lium
Dæmón-orops	Diác-rium	Dizygothè-ca
Dáh-lia	Dianél-la	Docýn-ia
Dà-is	Dián-thus	Dodecà-theon
Dalbér-gia	Diapén-sia	Dodonæ-a
Dà-lea	Diás-cia	Dolichán-dra
Dalechám-pia	Dicén-tra	Dól-ichos
Dalibár-da	Dichorisán-dra	Dolicothè-le
Dà-næ	Dicksò-nia	Dombè-ya
Dáph-ne	Dicranostíg-ma	Doò-dia
Daphniphýl-lum	Dictám-nus	Dór-itis

125

Dorón-icum
Dorotheán-thus
Dorstè-nia
Doryán-thes
Dorýc-nium
Doryóp-teris
Dossín-ia
Douglás-ia
Dovỳ-alis
Downín-gia
Doxán-tha
Drà-ba
Dracæ̀-na
Dracocéph-alum
Dracún-culus
Drì-mys
Drosán-themum
Drós-era
Dryán-dra
Drỳ-as
Dryóp-teris
Duchés-nea
Duggè-na
Durán-ta
Dù-rio
Duvà-lia
Dýck-ia
Dyschorís-te
Dysóx-ylum

Éb-enus
Ecbál-lium
Eccremocár-pus
Echevè-ria
Echidnóp-sis
Echinà-cea
Echinocác-tus
Echinocè-reus
Echinóch-loa
Echinocýs-tis
Echinomás-tus
Echinóp-anax
Echì-nops

Echinóp-sis
Echì-tes
Éch-ium, È-chium
Edgewór-thia
Edraián-thus
Ehrè-tia
Eichhór-nia
Elæág-nus
Elæ̀-is
Elæocár-pus
Elæodén-dron
Elaphoglós-sum
Elettà-ria
Eleusì-ne
Eliót-tia
Elodè-a
Elshólt-zia
Él-ymus
Embò-thrium
Emíl-ia
Emmenán-the
Emmenóp-terys
Ém-petrum
Encè-lia
Encephalár-tos
Enchylæ̀-na
Encýc-lia
Enkián-thus
Entelè-a
Enterolò-bium
Eomè-con
Ép-acris
Éph-edra
Epidén-drum
Epigæ̀-a
Epilò-bium
Epimè-dium
Epipác-tis
Epiphronì-tis
Epiphyllán-thus
Epiphýl-lum
Epís-cia
Epithelán-tha

Equisè-tum
Eragrós-tis
Erán-themum
Erán-this
Ercíl-la
Eremæ̀-a
Eremóch-loa
Eremocít-rus
Eremós-tachys
Eremù-rus
Erép-sia
È-ria
Erián-thus
Erì-ca
Ericamè-ria
Erigenì-a
Eríg-eron
Erì-nus
Eriobót-rya
Eriocéph-alus
Erióg-onum
Erióph-orum
Eriophýl-lum
Erióp-sis
Eriostè-mon
Eritrích-ium
Erlán-gea
Erò-dium
Erù-ca
Ervatà-mia
Erýn-gium
Erýs-imum
Erythè-a
Erythrì-na
Erythrò-nium
Erythróx-ylon
Escallò-nia
Eschschól-zia
Escobà-ria
Escón-tria
Euán-the
Eucalýp-tus
Eucharíd-ium

126

Eù-charis
Euchlǣ-na
Eù-comis
Eucóm-mia
Eucrýph-ia
Eugè-nia
Euón-ymus
Eupatò-rium
Euphór-bia
Euphò-ria
Eù-ploca
Euprítchár-dia
Euptè-lea
Eurò-tia
Eù-rya
Eurỳ-ale
Eù-scaphis
Eù-stoma
Eù-strephus
Eutáx-ia
Eutér-pe
Evò-dia
Evól-vulus
Éx-acum
Exóch-orda

Fabià-na
Fagopỳ-rum
Fà-gus
Fát-sia
Faucà-ria
Fè-dia
Feijò-a
Felíc-ia
Fenestrà-ria
Ferocác-tus
Ferò-nia
Feroniél-la
Fér-ula
Festù-ca
Fì-cus
Filipén-dula
Firmià-na

Fittò-nia
Fitzrò-ya
Flacoúrt-ia
Flemín-gia
Fœníc-ulum
Fontanè-sia
Forestiè-ra
Forsýth-ia
Fortunél-la
Forthergíl-la
Fouquiè-ria
Fragà-ria
Francò-a
Frankè-nia
Frasè-ra
Fráx-inus
Freè-sia
Fremón-tia
Freycinè-tia
Fritillà-ria
Frœlích-ia
Fù-chsia
Fumà-ria
Furcrǣ-a

Gà-gea
Gaillár-dia
Galactì-tes
Galán-thus
Gà-lax
Galeán-dra
Galè-ga
Gà-lium
Galtò-nia
Galvè-zia
Gamól-epis
Garbè-ria
Garcín-ia
Gardè-nia
Gár-rya
Gastè-ria
Gastrochì-lus
Gaulthè-ria

Gaù-ra
Gaús-sia
Gaỳ-a
Gaylussà-cia
Gazà-nia
Geitonoplè-sium
Gelsè-mium
Geniós-toma
Genì-pa
Genís-ta
Gentià-na
Geón-oma
Gerà-nium
Gerbè-ria
Gesnouín-ia
Gè-um
Gevuì-na
Gíl-ia
Gilibért-ia
Gillè-nia
Gínk-go
Gladì-olus
Glaucíd-ium
Glaúc-ium
Glaúx
Gledít-sia
Gliricíd-ia
Globulà-ria
Gloriò-sa
Glottiphýl-lum
Glycè-ria
Glycì-ne
Glycós-mis
Glycyrrhì-za
Glyptós-trobus
Gmelì-na
Gnaphà-lium
Godè-tia
Gomè-sa
Gomphocár-pus
Gompholò-bium
Gomphrè-na
Gongò-ra

127

Goò-dia
Gordò-nia
Gormà-nia
Gossýp-ium
Gourliè-a
Grabòw-skia
Grammatophýl-lum
Graptopét-alum
Graptophýl-lum
Gratì-ola
Greì-gia
Grevíl-lea
Grè-wia
Grè-yia
Grindè-lia
Griselín-ia
Guaì-cum
Guiliél-ma
Guizò-tia
Gunnè-ra
Guzmà-nia
Gymnocalýc-ium
Gymnóc-ladus
Gymnospò-ria
Gynandróp-sis
Gynè-rium
Gynù-ra
Gypsóph-ila

Habenà-ria
Habér-lea
Hacquè-tia
Hæmán-thus
Hæmà-ria
Hæmatóx-ylum
Hà-kea
Halè-sia
Halimodén-dron
Hamamè-lis
Hamatocác-tus
Hamè-lia
Harboù-ria
Hardenbér-gia

Harpephýl-lum
Harrís-ia
Hartwè-gia
Hatiò-ra
Hawór-thia
Hè-be
Hebenstreì-tia
Hedeò-ma
Héd-era
Hedycà-rya
Hedých-ium
Hedýs-arum
Hedyscè-pe
Heì-mia
Helè-nium
Heliám-phora
Helianthél-la
Helián-themum
Helián-thus
Helichrỳ-sum
Helicodíc-eros
Helicò-nia
Heliocè-reus
Helióp-sis
Heliotrò-pium
Helíp-terum
Helléb-orus
Helò-nias
Helwín-gia
Helxì-ne
Hemerocál-lis
Hemián-dra
Hemicỳ-clia
Hemíg-raphis
Hemionì-tis
Hemiptè-lia
Hepát-ica
Heraclè-um
Hererò-a
Hernià-ria
Hesperà-loe
Hesperethù-sa
Hés-peris

Hesperoyúc-ca
Heterocén-tron
Heteromè-les
Heterós-pathe
Heterospér-mum
Heuchè-ra
Hè-vea
Hibbér-tia
Hibís-cus
Hicksbeà-chia
Hidalgò-a
Hierà-cium
Hippeàs-trum
Hippocrè-pis
Hippóph-aë
Hoffmán-nia
Hohè-ria
Hól-cus
Holmskiól-dia
Holodís-cus
Holoptè-lea
Homalán-thus
Homalocéph-ala
Homaloclà-dium
Homalomè-na
Hór-deum
Hormì-num
Hosáck-ia
Hò-sta
Houllè-tia
Houstò-nia
Houttuỳ-nia
Hò-vea
Hovè-nia
Hòw-ea
Hoỳ-a
Huér-nia
Hufelán-dia
Humà-ta
Hù-mea
Hù-mulus
Hunnemán-nia
Hù-ra

Hutchín-sia
Hyacín-thus
Hydrán-gea
Hydrás-tis
Hydriastè-le
Hydróch-aris
Hydrò-cleys
Hydrocót-yle
Hydrò-lea
Hydrophýl-lum
Hydrós-me
Hylocè-reus
Hymenǽ-a
Hymenán-thera
Hymenocál-lis
Hymenós-porum
Hyophór-be
Hyoscỳ-amus
Hypér-icum, Hy-
  perì-cum
Hyphǽ-ne
Hypocalým-ma
Hypochœ-ris
Hypól-epis
Hypóx-is
Hyssò-pus
Hýs-trix

Ibè-ris
Ibò-za
Idè-sia
Íd-ria
Ì-lex
Illíc-ium
Impà-tiens
Incarvíl-lea
Indigóf-era
Ín-ga
Ingenhoù-zia
Ín-ula
Iochrò-ma
Ioníd-ium
Ionopsíd-ium

Ipomœ-a
Iresì-ne
Ì-ris
Ís-atis
Isér-tia
Isolò-ma
Isopléx-is
Isopò-gon
Isopỳ-rum
Isót-oma
Ít-ea
Íx-ia
Ixiolír-ion
Ixò-ra

Jacarán-da
Jacobín-ia
Jacquemón-tia
Jasiò-ne
Jás-minum
Ját-ropha
Jeffersò-nia
Jovellà-na
Juà-nia
Jubǽ-a
Júg-lans
Jún-cus
Juníp-erus
Jussiǽ-a
Justíc-ia

Kadsù-ra
Kagenéck-ia
Kalán-choë
Kál-mia
Kennĕd-ia
Kén-tia
Kentióp-sis
Kernè-ra
Kér-ria
Keteleè-ria
Kíck-xia

Kigè-lia
Kirengeshò-ma
Kitaibè-lia
Knì-ghtia
Kniphò-fia
Kò-chia
Koelè-ria
Kœlreutè-ria
Kò-kia
Kolkwít-zia
Korthál-sia
Kostelét-zkya
Kramè-ria
Kríg-ia
Kù-hnia
Kún-zea

Labúr-num
Lachenà-lia
Lactù-ca
Lǽ-lia
Læliocátt-leya
Lagenà-ria
Lagerstrœ-mia
Lagunà-ria
Lagù-rus
Lallemán-tia
Lamár-ckia
Lambért-ia
Là-mium
Lampᵣán-thus
Lantà-na
Lapagè-ria
Lapeiroù-sia
Láp-pula
Lardizabà-la
Là-rix
Lár-rea
Laserpít-ium
Lasthè-nia
Latà-nia
Láth-yrus
Laurè-lia

Laù-rus
Laván-dula
Laván-ga
Lavát-era
Lawsò-nia
Là-yia
Lè-dum
Leè-a
Leiophýl-lum
Lemaireocè-reus
Lém-na
Léns
Leonò-tis
Leontopò-dium
Leonù-rus
Lép-achys
Lepíd-ium
Leptóch-loa
Leptodác-tylon
Leptodér-mis
Leptóp-teris
Leptopỳ-rum
Leptospér-mum
Leptós-yne
Lép-totes
Leschenaù-ltia
Lespedè-za
Lesquerél-la
Lettsò-mia
Leucadén-dron
Leucæ̀-na
Leuchè-ria
Leucóc-rinum
Leucò-jum
Leucophýl-lum
Leucóth-oë
Leù-zea
Levís-ticum
Lewís-ia
Leycestè-ria
Lià-tris
Libér-tia
Libocéd-rus

Licuà-la
Ligulà-ria
Ligús-ticum
Ligù-strum
Líl-ium
Limnán-thes
Limnóch-aris
Limò-nium
Linán-thus
Linà-ria
Lindelò-fia
Linnæ̀-a
Linospà-dix
Linós-yris
Lì-num
Líp-aris
Líp-pia
Liquidám-bar
Liriodén-dron
Lirì-ope
Listè-ra
Lì-tchi
Lithocár-pus
Lithodò-ra
Lithofrág-ma
Líth-ops
Lithospér-mum
Lithræ̀-a
Lít-sea
Livistò-na
Loà-sa
Lobè-lia
Lobív-ia
Lobulà-ria
Lockhár-tia
Lodoì-cea
Loesè-lia
Logà-nia
Loiseleù-ria
Lò-lium
Lomà-tia
Lomà-tium
Lò-nas

Lonchocár-pus
Loníc-era
Lopè-zia
Lophóph-ora
Loropét-alum
Lò-tus
Lucù-lia
Lucù-ma
Ludwíg-ia
Luét-kea
Lúf-fa
Lunà-ria
Lupì-nus
Lycás-te
Lých-nis
Lýc-ium
Lycopér-sicon
Lycopò-dium
Lýc-opus
Lýc-oris
Lygò-dium
Lyò-nia
Lyonothám-nus
Lysichì-tum
Lysimà-chia
Lýth-rum

Maà-ckia
Mà-ba
Machæroсè-reus
Mackà-ya
Macleà-ya
Maclù-ra
Macradè-nia
Macróp-iper
Macrozà-mia
Madacà-mia
Maddè-nia
Mà-dia
Mæ̀-sa
Magnò-lia
Mahér-nia
Mahobér-beris

Mahò-nia
Maián-themum
Majorà-na
Malách-ra
Malacocár-pus
Malacóth-rix
Malcò-mia
Maléph-ora
Mallò-tus
Mál-ope
Malortì-ea
Malpíg-hia
Mál-va
Malvás-trum
Malvavís-cus
Mamillóp-sis
Mám-mea
Mammillà-ria
Mandevíl-la
Mandrág-ora
Manét-tia
Manfrè-da
Mangíf-era
Mán-ihot
Manulè-a
Marán-ta
Marát-tia
Margyricár-pus
Már-ica
Marrù-bium
Marsdè-nia
Marsíl-ea
Martinè-zia
Mascarenhà-sia
Masdevál-lia
Mathì-ola
Matricà-ria
Maurán-dia
Maxillà-ria
Maytè-nus
Mà-zus
Meconóp-sis
Medè-ola

Medicà-go
Mediníl-la
Mediocác-tus
Melaleù-ca
Melampò-dium
Melán-thium
Melasphæ̀-rula
Melás-toma
Mè-lia
Melián-thus
Mél-ica
Melicóc-ca
Melicỳ-tus
Melilò-tus
Meliós-ma
Melís-sa
Melít-tis
Melocác-tus
Melò-thria
Menispér-mum
Menodò-ra
Mén-tha
Mentzè-lia
Menyán-thes
Menziè-sia
Merà-tia
Mercurià-lis
Mertén-sia
Mér-yta
Mesembryán-
   themum
Més-pilus
Metrosidè-ros
Mè-um
Michaù-xia
Michè-lia
Micò-nia
Microcít-rus
Microcỳ-cas
Microglós-sa
Microlè-pia
Micromè-ria
Micrós-tylis

Mikà-nia
Míl-la
Miltò-nia
Mimò-sa
Mím-ulus
Mím-usops
Miráb-ilis
Miscán-thus
Mitchél-la
Mitél-la
Mitrà-ria
Molín-ia
Molopospér-mum
Mól-tkia
Molucél-la
Momór-dica
Monár-da
Monardél-la
Món-do
Monè-ses
Monotág-ma
Monót-ropa
Monstè-ra
Montanò-a
Montezù-ma
Món-tia
Monvíl-lea
Morè-a
Morì-na
Morín-da
Morín-ga
Mò-rus
Moschà-ria
Mucù-na
Murræ̀-a
Mù-sa
Muscà-ri
Mutís-ia
Myóp-orum
Myosotíd-eum
Myosò-tis
Myrì-ca
Myricà-ria

Myriocéph-alus
Myriophýl-lum
Myrospér-mum
Myróx-ylon
Myrrhì-num
Mýr-rhis
Mýr-sine
Myrtillocác-tus
Mýr-tus
Mystacíd-ium

Nægè-lia
Nanán-thus
Nandì-na
Nán-norrhops
Narcís-sus
Nastúr-tium
Navarrét-ia
Neíl-lia
Nelúm-bium
Nemás-tylis
Nemè-sia
Nemopán-thus
Nemóph-ila
Neobés-seya
Neolloỳ-dia
Nepén-thes
Nép-eta
Nephról-epis
Nerì-ne
Nè-rium
Nertè-ra
Neviù-sia
Neyraù-dia
Nicán-dra
Nicotià-na
Nidulà-rium
Nierembér-gia
Nigél-la
Nì-pa
Nolà-na
Nolì-na
Nól-tea

Nopà-lea
Nopalxò-chia
Normán-bya
Nothofà-gus
Nothól-cus
Nothóp-anax
Nothoscór-dum
Notò-nia
Nototrích-ium
Nyctán-thes
Nyctocè-reus
Nymphǽ-a
Nymphoì-des
Nymphozán-thus
Nýs-sa

Óch-na
Ò-cimum
Octomè-ria
Odontiò-da
Odontoglós-sum
Odontonè-ma
Odontosò-ria
Œnothè-ra
Ò-lea
Oleà-ria
Oliverán-thus
Omphalò-des
Oncíd-ium
Ón-coba
Onób-rychis
Onoclè-a
Onò-nis
Onorpór-dum
Onós-ma
Onosmò-dium
Oných-ium
Ophioglós-sum
Ò-phrys
Oplís-menus
Opún-tia
Ór-chis
Oreóp-anax

Oríg-anum
Oríx-a
Ormò-sia
Ornithíd-ium
Ornithochì-lus
Ornithóg-alum
Orníth-opus
Orón-tium
Oróx-ylon
Orthocár-pus
Orỳ-za
Osculà-ria
Osmán-thus
Osmarò-nia
Osmorhì-za
Osmún-da
Osetomè-les
Ostròw-skia
Ós-trya
Othón-na
Ourís-ia
Óx-alis
Oxè-ra
Oxydén-drum
Oxylò-bium
Oxypét-alum
Oxýt-ropis

Pachì-ra
Pachís-tima
Pachycè-reus
Pachýph-ytum
Pachyrhì-zus
Pachysán-dra
Pachýs-tachys
Pachystè-gia
Pæò-nia
Palà-quium
Palicoù-rea
Palisò-ta
Paliù-rus
Palmerél-la
Pà-nax

Pancrà-tium
Pandà-nus
Pandò-rea
Pán-icum
Papà-ver
Paphiopè-dilum
Paradì-sea
Paramíg-nya
Parietà-ria
Pà-ris
Parkinsò-nia
Parmentiè-ra
Parnás-sia
Paróch-etus
Paroných-ia
Parrò-tia
Parrotióp-sis
Parthè-nium
Parthenocís-sus
Pás-palum
Passiflò-ra
Pastinà-ca
Paullín-ia
Paulòw-nia
Pavò-nia
Pediculà-ris
Pedilán-thus
Pediocác-tus
Pelargò-nium
Pelecýph-ora
Pellæ̀-a
Pelliò-nia
Peltán-dra
Peltà-ria
Peltiphýl-lum
Peltóph-orum
Peniocè-reus
Pennán-tia
Pennisè-tum
Penstè-mon
Pentaglót-tis
Pentapterýg-ium
Peperò-mia

Perés-kia
Pereskióp-sis
Perè-zia
Períl-la
Períp-loca
Peristè-ria
Perís-trophe
Pernét-tia
Peróv-skia
Pér-sea
Persoò-nia
Pescatò-ria
Petalostè-mum
Petasì-tes
Petivè-ria
Petrè-a
Petrocál-lis
Petrocóp-tis
Petróph-ila
Petróph-ytum
Petroselì-num
Pettè-ria
Pctù-nia
Peucéd-anum
Peù-mus
Phacè-lia
Phædrán-thus
Phà-ius
Phalænóp-sis
Phál-aris
Phasè-olus
Phebà-lium
Phellodén-dron
Phellospér-ma
Philadél-phus
Philè-sia
Philibér-tia
Phillýr-ea
Philodén-dron
Phlè-um
Phlò-mis
Phlóx
Phœ̀-be

Phœ̀-nix
Pholidò-ta
Phór-mium
Photín-ia
Phragmì-tes
Phygè-lius
Phýl-ica
Phyllág-athis
Phyllán-thus
Phyllì-tis
Phyllocác-tus
Phyllóc-ladus
Phyllód-oce
Phyllós-tachys
Phýs-alis
Physocár-pus
Physosì-phon
Physostè-gia
Phytél-ephas
Phyteù-ma
Phytolác-ca
Pì-cca, Píc-ca
Píc-ris
Pì-eris
Píl-ea, Pì-lea
Pilocè-reus
Pimè-lea
Pimén-ta
Pimpinél-la
Pinán-ga
Pinguíc-ula
Pì-nus
Pì-per
Piptadè-nia
Piptán-thus
Piquè-ria
Pisò-nia
Pistà-cia
Pís-tia
Pì-sum
Pitcaír-nia
Pithecellò-bium
Pithecoctè-nium

Pittós-porum
Pityrográm-ma
Plagián-thus
Planè-ra
Plantà-go
Plát-anus
Platycà-rya
Platycè-rium
Platycò-don
Platymís-cium
Platystè-mon
Pleiogýn-ium
Pleiò-ne
Pleiospì-los
Pleurothál-lis
Plumbà-go
Plumè-ria
Pò-a
Podachæ̀-nium
Podalýr-ia
Podocár-pus
Podól-epis
Podophýl-lum
Pogò-nia
Poincià-na
Polanís-ia
Polemò-nium
Polián-thes
Poliothýr-sis
Pól-lia
Polyandrocóc-cos
Polýg-ala
Polygón-atum
Polýg-onum
Polypò-dium
Polypò-gon
Polýp-teris
Polýs-cias
Polystà-chya
Polýs-tichum
Pomadér-ris
Poncì-rus
Pongà-mia

Pontadè-ria
Póp-ulus
Porà-na
Portlán-dia
Portulà-ca
Portulacà-ria
Posoquè-ria
Potentíl-la
Potè-rium
Pò-thos
Prà-tia
Prém-na
Prenán-thes
Prím-ula
Prinsè-pia
Pritchár-dia
Proboscíd-ea
Promenæ̀-a
Prosò-pis
Prostanthè-ra
Prò-tea
Prunél-la
Prù-nus
Pseuderán-themum
Pseudolà-rix
Pseudóp-anax
Pseudophœ̀-nix
Pseudotsù-ga
Psíd-ium
Psophocár-pus
Psorà-lea
Psychò-tria
Ptè-lea
Pterè-tis
Pteríd-ium
Ptè-ris
Pterocà-rya
Pterocéph-alus
Pterospér-mum
Pterós-tyrax
Pterygò-ta
Ptychorà-phis
Ptyschospér-ma

Puerà-ria
Pulicà-ria
Pulmonà-ria
Pultenæ̀-a
Pù-nica
Púr-shia
Puschkín-ia
Pù-ya
Pycnán-themum
Pychnós-tachys
Pyracán-tha
Pyrè-thrum
Pýr-ola
Pyrostè-gia
Pỳ-rus
Pyxidanthè-ra

Quám-oclit
Quás-sia
Quér-cus
Quillà-ja
Quín-cula
Quintín-ia
Quisquà-lis

Radermách-ia
Rajà-nia
Ramón-da
Ranè-vea
Ranún-culus
Raoù-lia
Ráph-anus
Ráph-ia
Raphiól-epis
Rathbù-nia
Ravenà-la
Rebù-tia
Rehmán-nia
Reichár-dia
Reinéck-ia
Reinwár-dtia
Renanthè-ra
Resè-da

Rhabdothám-nus
Rhagò-dia
Rhám-nus
Rhaphithám-nus
Rhapidophýl-lum
Rhà-pis
Rhektophýl-lum
Rhè-um
Rhéx-ia
Rhinán-thus
Rhipóg-onum
Rhíp-salis
Rhizóph-ora
Rhodóch-iton
Rhododén-dron
Rhodomýr-tus
Rhodós-tachys
Rhodothám-nus
Rhodót-ypus
Rhœ̀-o
Rhombophýl-lum
Rhopalós-tylis
Rhús
Rhynchò-sia
Rhynchós-tylis
Rhyticò-cos
Rì-bes
Ríc-cia
Richár-dia
Ríc-inus
Ricò-tia
Rivì-na
Robín-ia
Rò-chea
Rodgér-sia
Rodriguè-zia
Roemè-ria
Ròh-dea
Rollín-ia
Romanzóf-fia
Rondelè-tia
Rò-sa
Roschè-ria

Roseocác-tus
Rosmarì-nus
Roúp-ala
Royè-na
Roystò-nea
Rùbia
Rù-bus
Rudbéck-ia
Ruél-lia
Rù-mex
Rús-cus
Russè-lia
Rù-ta

Sà-bal
Sác-charum
Sadlè-ria
Sagerè-tia
Sagì-na
Sagittà-ria
Saintpaù-lia
Salicór-nia
Sà-lix
Salpichrò-a
Salpiglós-sis
Sál-sola
Sál-via
Salvín-ia
Samanè-a
Sambù-cus
Sám-olus
Samuè-la
Sanchè-zia
Sanguinà-ria
Sanguisór-ba
Sanseviè-ria
Sán-talum
Santolì-na
Sanvità-lia
Sapín-dus
Sà-pium
Saponà-ria
Sapò-ta

Sarà-ca
Sarcán-thus
Sarchochì-lus
Sarcocóc-ca
Sarcoglót-tis
Sarracè-nia
Sás-a
Sás-safras
Saturè-ja
Sauróm-atum
Saurù rus
Saussù-rea
Saxegothæ̀-a
Saxíf-raga
Scabiò-sa
Scelè-tium
Schauè-ria
Scheè-lea
Schefflè-ra
Schì-ma
Schì-nus
Schisán-dra
Schismatoglót-tis
Schiveréck-ia
Schizæ̀-a
Schizán-thus
Schizobasóp-sis
Schizocén-tron
Schizocò-don
Schizolò-bium
Schizopét-alon
Schizophrág-ma
Schizós-tylis
Schlumbergè-ra
Schombúrg-kia
Schò-tia
Schrán-kia
Sciadóp-itys
Scíl-la
Scindáp-sus
Scír-pus
Sclerocác-tus
Scleropò-a

Scól-ymus
Scorpiù-rus
Scorzonè-ra
Scrophulà-ria
Scutellà-ria
Scuticà-ria
Secà-le
Sè-chium
Securíd-aca
Securíg-era
Sè-dum
Selaginél-la
Selenicè-reus
Selenipè-dium
Sém-ele
Semmán-the
Sempervì-vum
Senè-cio
Sequò-ia
Serenò-a
Sericocár-pus
Serís-sa
Serjà-nia
Serrát-ula
Sés-amum
Sesbà-nia
Setà-ria
Severín-ia
Shephér-dia
Shór-tia
Sibiræ-a
Sibthór-pia
Sicà-na
Síc-yos
Sidál-cea
Siderì-tis
Sideróx-ylon
Sigmatós-talix
Silè-ne
Síl-phium
Síl-ybum
Simmónd-sia
Sinnín-gia

Sinomè-nium
Siphonán-thus
Sisyrín-chium
Sì-um
Skím-mia
Smilacì-na
Smì-lax
Sobrà-lia
Solán-dra
Solà-num
Soldanél-la
Solidà-go
Solís-ia
Sól-lya
Sǒn-chus
Sonerì-la
Sóph-ora, Sophò-ra
Sophronì-tis
Sorbà-ria
Sorbarò-nia
Sór-bus
Sparáx-is
Sparmán-nia
Spár-tium
Spathiphýl-lum
Spathò-dea
Spathoglót-tis
Speculà-ria
Spér-gula
Sphác-ele
Sphærál-cea
Spigè-lia
Spilán-thes
Spinà-cia
Spiræ-a
Spirán-thes
Spironè-ma
Spón-dias
Sprà-guea
Sprekè-lia
Spyríd-ium
Stà-chys
Stachytarphè-ta

Stachyù-rus
Stanhò-pea
Stapè-lia
Staphylè-a
Stát-ice
Stauntò-nia
Steironè-ma
Stellà-ria
Stenán-driùm
Stenán-thium
Stenocár-pus
Stenochlæ̀-na
Stenoglót-tis
Stenolò-bium
Stenospermà-tion
Stenotáph-rum
Stephanán-dra
Stephanomè-ria
Stephanò-tis
Stercù-lia
Sterlít-zia
Sternbér-gia
Stevensò-nia
Stè-via
Stewár-tia
Stigmaphýl-lon
Stilbocár-pa
Stì-pa
Stizolò-bium
Stokè-sia
Stranvæ̀-sia
Stratiò-tes
Streptocár-pus
Strelít-zia
Streptocár-pus
Strép-topus
Streptosò-len
Strobilán-thes
Stromán-the
Strombocác-tus
Strombocár-pa
Strých-nos
Stylíd-ium

Stylóph-orum
Stylophýl-lum
Stỳ-rax
Succì-sa
Sutherlán-dia
Suttò-nia
Swainsò-na
Swietè-nia
Swinglè-a
Symphoricár-pos
Symphyán-dra
Sým-phytum
Symplocár-pus
Sým-plocos
Synadè-nium
Syncár-pia
Syncchán-thus
Syntherís-ma
Sýn-thyris
Syrín-ga

Tabcbù-ia
Tabernæmontà-na
Tác-ca
Tæníd-ia
Tagè-tes
Taiwà-nia
Talì-num
Tamarín-dus
Tám-arix
Tà-mus
Tanacè-tum
Taraktogè-nos
Taráx-acum
Taxò-dium
Táx-us
Téc-oma
Tecomà-ria
Téc-tona
Telè-phium
Tellì-ma
Telò-pea

Templetò-nia
Tephrò-sia
Terminà-lia
Ternstrœ-mia
Testudinà-ria
Tetracén-tron
Tetraclì-nis
Tetragò-nia
Tetráp-anax
Tetrapathǽ-a
Tetrathè-ca
Teù-crium
Thà-lia
Thalíc-trum
Thamnocál-amus
Thè-a
Thelcspér-ma
Thelocác-tus
Thelypò-dium
Theobrò-ma
Thermóp-sis
Thespè-sia
Thevè-tia
Thlás-pi
Thomás-ia
Thrì-nax
Thrixspér-mum
Thryál-lis
Thù-ja
Thujóp-sis
Thunbér-gia
Thù-nia
Thỳ-mus
Thysanolǽ-na
Thysanò-tus
Tiarél-la
Tibouchì-na
Tigríd-ia
Tíl-ia
Tilländ-sia
Tinán-tia
Tipuà-na
Titanóp-sis

Tithò-nia
Tocò-ca
Tolmiè-a
Tól-pis
Torè-nia
Torrè-ya
Tovà-ra
Townsén-dia
Trachè-lium
Trachelospér-mum
Trachycár-pus
Trachým-ene
Trachystè-mon
Tradescán-tia
Tragopò-gon
Trà-pa
Trautvettè-ria
Trè-ma
Trevè-sia
Trevò-a
Tricalýs-ia
Trichíl-ia
Trichocè-reus
Trichodiadè-ma
Tricholǽ na
Trichopíl-ia
Trichosán-thes
Trichós-porum
Trichostè-ma
Tricýr-tis
Trì-dax
Trientà-lis
Trifò-lium
Trigonél-la
Tríl-isa
Tríl-lium
Trimè-za
Triós-teum
Triphà-sia
Tríp-laris
Tripterýg-ium
Trisè-tum
Tristà-nia

Trithrì-nax
Trít-icum
Tritò-nia
Trochodén-dron
Tról-lius
Tropæ-olum
Tsù-ga
Tù-lipa
Tù-nica
Tupidán-thus
Turræ-a
Tussà-cia
Tussilà-go
Tỳ-pha

Ù-lex
Úll-ucus
Úl-mus
Umbellulà-ria
Ungnà-dia
Unì-ola
Urbín-ia
Ù-rera
Urgín-ea
Uropáp-pus
Ursín-ia
Urtì-ca
Utriculà-ria
Uvulà-ria

Vaccín-ium
Valerià-na
Valerianél-la
Vallà-ris
Vallisnè-ria
Vallò-ta
Vancouvè-ria
Ván-da
Vandóp-sis
Vaniè-ria

Vaníl-la
Veì-tchia
Veltheì-mia
Veníd-ium
Vè-pris
Verà-trum
Verbás-cum
Verbè-na
Verbesì-na
Vernò-nia
Verón-ica
Veronicás-trum
Verschaffél-tia
Verticór-dia
Vesicà-ria
Vibúr-num
Víc-ia
Victò-ria
Víg-na
Villarè-sia
Vín-ca
Vincetóx-icum
Vì-ola
Virgíl-ia
Vì-tex
Vì-tis
Vittadín-ia
Vriè-sia

Wahlenbér-gia
Waldsteì-nia
Wallích-ia
Walthè-ria
Warscewiczél-la
Warszewíc-zia
Washingtò-nia
Watsò-nia
Wedè-lia
Weigè-la
Weinmán-nia
Wérck-lea

Westríng-ia
Widdringtò-nia
Wigán-dia
Wilcóx-ia
Wistè-ria
Woód-sia
Woodwár-dia
Wulfè-nia
Wyè-thia

Xanthís-ma
Xanthóc-eras
Xanthorrhœ-a
Xanthosò-ma
Xerán-themum
Xerophýl-lum
Xylò-bium
Xylophýl-la

Yúc-ca

Zaluzián-skya
Zà-mia
Zantedés-chia
Zanthorhì-za
Zanthóx-ylum
Zauschnè-ria
Zè-a
Zebrì-na
Zelkò-va
Zenò-bia
Zephyrán-thes
Zín-giber
Zín-nia
Zizà-nia
Zíz-yphus
Zoý-sia
Zygád-enus
Zygocác-tus
Zygopét-alum

LIST II

Specific or trivial Latin names, with spelling, indication of pronunciation and suggestion of botanical application.

Grave accent (`` ` ``), long vowel;
acute accent (´), short or
other vowel sounds

abbrevià-tus: abbreviated, shortened
abietì-nus: abies-like
abortì-vus: aborted, parts failing
abrotanifò-lius: abrotanum-leaved
abrúp-tus: abrupt
absinthoì-des: absinthe-like
abyssín-icus: Abyssinian
acanthifò-lius: acanthus-leaved
acanthóc-omus: spiny-haired or -crowned
acaù-lis: stemless
ác-colus: dwells near
acéph-alus: headless
acér-bus: harsh or sour
acerifò-lius: maple-leaved
aceroì-des: maple-like
acerò-sus: needle-shaped
achilleæfò-lius: achillea-leaved
aciculà-ris: needle-like
acidís-simus: exceedingly sour
ác-idus: acid, sour
acinà-ceus: scimitar- or saber-shaped
acinacifò-lius: scimitar-leaved
acinacifór-mis: scimitar-shaped
aconitifò-lius: aconite-leaved
à-cris: acrid, sharp

acrostichoì-des: acrostichum-like
acrót-riche: hairy-lipped
aculeatís-simus: very prickly
aculeà-tus: prickly
acuminatifò-lius: acuminate-leaved
acuminatís-simus: very acuminate
acuminà tus: acuminate, long pointed, tapering
acutàn-gulus: acutely angled
acutíf-idus: acutely cut
acutifò-lius: acutely leaved, sharp-leaved
acutíl-obus: acutely lobed
acutipét-alus: petals acute
acutís-simus: very acute
acù-tus: acute, sharp-pointed
adenóph-orus: gland-bearing
adenophýl-lus: glandular-leaved
adenóp-odus: glandular-footed
adiantoì-des: adiantum-like
admiráb-ilis: admirable, noteworthy
adnà-tus: adnate, joined to
adonidifò-lius: adonis-leaved
adprés-sus: pressed against
adscén-dens: ascending
adsúr-gens: ascending
adún-cus: hooked

139

ád-venus: newly arrived, adventive

ægyptì-acus: Egyptian

ǽm-ulus: emulative, imitating

æquinoctià-lis: pertaining to the equinox, mid-tropical

æquipét-alus: equal-petaled

æquitríl-obus: equally three-lobed

aè-rius: aërial

æruginò-sus: rusty, rust-colored

æstivà-lis: pertaining to summer

æstì-vus: summer

æthióp-icus: Ethiopian, African

affì-nis: related

à-fra: African

africà-nus: African

agavoì-des: agave-like

ageratifò-lius: ageratum-leaved

ageratoì-des: ageratum-like

aggregà-tus: aggregate, clustered

agrà-rius: of the fields

agrés-tis: of or pertaining to the fields

agrifò-lius: scabby-leaved

aizoì-des: aizoon-like

alà-tus: winged

albés-cens: whitish, becoming white

ál-bicans: whitish

albicaù-lis: white-stemmed

ál-bidus: white

albiflò-rus: white-flowered

ál-bifrons: white-fronded

albispì-nus: white-spined

albocínc-tus: white-girdled, white-crowned

albo-píc-tus: white-painted

albo-pilò-sus: white-shaggy

albospì-cus: white-spiked

ál-bulus: whitish

ál-bus: white

alchemilloì-des: alchemilla-like

alcicór-nis: elk-horned

alép-picus: of Aleppo (Syria)

alexandrì-nus: of Alexandria (Egypt)

ál-gidus: cold

aliè-nus: foreign

allià-ceus: of the alliums, garlic-like

alliariæfò-lius: alliaria-leaved

alnifò-lius: alder-leaved

aloì-des, alooì-des: aloe-like

aloifò-lius: aloe-leaved

alopecurioì-des: alopecurus-like

alpés-tris: nearly alpine

alpíg-enus: alpine

alpì-nus: alpine

altà-icus: of the Altai Mountains (Siberia)

altér-nans: alternating

alternifò-lius: alternate-leaved

altér-nus: alternating, alternate

althæoi-des: althæa-like, hollyhock-like

ál-tifrons: tall-fronded

altís-simus: very tall, tallest

ál-tus: tall

alúm-nus: well nourished, flourishing, strong

alyssoì-des: alyssum-like

amáb-ilis: lovely

amaranthoì-des: amaranth-like

amarantíc-olor: amaranth-colored

amaricaù-lis: bitter-stemmed

amà-rus: bitter

amazón-icus: of the River Amazon region

ambíg-uus: ambiguous, doubtful

amblỳ-odon: blunt-toothed

ambrosioì-des: ambrosia-like

amelloì-des: amellus-like

americà-nus: American

amethýs-tinus: amethystine, violet-colored

amethystoglós-sus: amethyst-tongued

ammóph-ilus: sand-loving

amœ-nus: charming, pleasing

amphíb-ius: amphibious, growing on land or in water

amplexicaù-lis: stem-clasping

amplexifò-lius: leaf-clasping

amplià-tus: enlarged

amplís-simus: most or very ample

úm-plus: ample, large

amurén-sis: of the Amur River region (northeastern Asia)

amygdalifór-mis: almond-shaped

amygdál-inus: almond-like

amygdaloì-des: almond-like

anacán-thus: without spines

anacardioì-des: anacardium-like

anagyroì-des: anagyris-like

anatól-icus: of Anatolia (Asia Minor)

án-ceps: two-headed, two-edged

andíc-olus: native of the Andes

andì-nus: Andine, pertaining to the Andes

andróg-ynus: hermaphrodite

androsà-ceus: like androsace

androsæmifò-lius: androsæmum-leaved

anemoneflò-rus: anemone-flowered

anemonefò-lius, anemonifò-lius: anemone-leaved

anemonoì-des: anemone-like

anethifò-lius: anethum-leaved

aneù-rus: nerveless

anfractuò-sus: twisted

án-glicus: English, of England

anguì-nus: snaky, snake-like

angulà-ris, angulà-tus: angular, angled

angulò-sus: angled, full of corners

angustifò-lius: narrow-leaved

angús-tus: narrow

anisà-tum: anise-scented

anisodò-rus: anise-odor

anisophýl-lus: unequal-leaved

annót-inus: year-old

annulà-ris: annular, ringed

annulà-tus: annular

án-nuus: annual

anóm-alus: anomalous, out of the ordinary or usual

anopét-alus: erect-petaled

antárc-ticus: of the Antarctic regions

anthemoì-des: anthemis-like

anthocrè-ne: flower-fountain

anthyllidifò-lius: anthyllis-leaved

antillà-ris: of the Antilles (West Indies)

antíp-odum: of the antipodes

antiquò-rum: of the ancients

antì-quus: ancient

antirrhiniflò-rus: antirrhinum-flowered

antirrhinoì-des: antirrhinum-like, snapdragon-like

apennì-nus: pertaining to the Apennines (Italy)

apér-tus: uncovered, bare, open

apét-alus: without petals

aphýl-lus: leafless

apiculà-tus: apiculate, tipped with a point

apíf-era: bee-bearing

apiifò-lius: apium-leaved

áp-odus: footless

apopét-alus: having free petals

appendiculà-tus: appendaged

applanà-tus: flattened

141

applicà-tus: joined, attached
áp-ricus: uncovered
áp-terus: wingless
aquát-icus, aquát-ilis: aquatic
à-queus: aqueous, watery
aquilegifò-lius: aquilegia-leaved
aquilì-nus: aquiline, eagle-like
aráb-icus: Arabian
arachnoì-des: spider-like, cob-
webby
araliæfò-lius: aralia-leaved
arborés-cens: becoming tree-
like, woody
arbò-reus: tree-like
arbús-culus: like a small tree
arbutifò-lius: arbutus-leaved
árc-ticus: arctic
arenà-rius, arenò-sus: of sand or
sandy places
areolà-tus: pitted
argentà-tus: silvery, silvered
argenteo-guttà-tus: silver-
spotted
argén-teus: silvery
argillà-ceus: of clay
argophýl-lus: silver-leaved
argù-tus: sharp-toothed
argyræ-us: silvery
argyróc-omus: silver-haired
argyroneù-rus: silver-nerved
argyrophýl-lus: silver-leaved
ár-idus: arid
arietì-nus: like a ram's head
aristà-tus: aristate, bearded
aristò-sus: bearded
arizón-icus: of Arizona
arkansà-nus: of Arkansas
armà-tus: armed
armillà-ris: with a bracelet, arm-
ring, or collar
aromát-icus: aromatic
arréct-us: raised up, erect
artemisioì-des: artemisia-like

articulà-tus: articulated, jointed
arundinà-ceus: reed-like
arvén-sis: pertaining to cultivated fields
asarifò-lius: asarum-leaved
ascalón-icus: of Ascalon (Syria)
ascén-dens: ascending
asclepiadè-us: asclepias-like
asiát-icus: Asian
ás-per: rough
asperà-tus: rough
aspericaù-lis: rough-stemmed
asperifò-lius: rough-leaved
aspér-rimus: very rough
asphodeloì-des: asphodelus-like
asplenifò-lius: asplenium-leaved
assím-ilis: similar, like to
assúr-gens: ascending
assurgentiflò-rus: flowers ascending
asteroì-des: aster-like
astù-ricus: of Asturia, Spain
à-ter: coal-black
atlán-ticus: Atlantic
atomà-rius: speckled
atrà-tus: blackened
atriplicifò-lius: atriplex-leaved,
orach-leaved
atrocár-pus: dark-fruited
atropurpù-reus: dark purple
atrór-ubens: dark red
atrosanguín-eus: dark blood-red
atroviolà-ceus: dark violet
atróv-irens: dark green
attenuà-tus: attenuated, produced to a point
át-ticus: pertaining to Attica or
Athens, Greece
aubretioì-des: aubretia-like
augustís-simus: very notable
augús-tus: august, notable, majestic
aurantì-acus: orange-red

142

LIST II. SPECIFIC NAMES

aurantifò-lius: golden-leaved
aurè-olus: golden
aù-reus: golden
auriculà-tus: eared
auríc-omus: golden-haired
aurì-tus: eared
australién-sis: belonging to Australia
austrà-lis: southern
austrì-acus: Austrian
austrì-nus: southern
autumnà-lis: autumnal
avicenniæfò-lius: avicennia-leaved
aviculà-ris: pertaining to birds
à-vium: of the birds
axillà-ris: axillary
azaleoì-des: azalea-like
azór-icus: of the Azores
azù-reus: azure, sky-blue

babylón-icus: Babylonian
bác-cans, baccà-tus: berried
baccíf-era: berry-bearing
bacterióph-ilus: bacteria-loving
baleár-icus: of the Balearic Islands
balsà-meus: balsamic
balsamíf-era: balsam-bearing
bál-ticus: of the Baltic
bambusoì-des: bamboo-like
banát-icus: of Banat (Hungary)
bár-barus: foreign
barbát-ulus: somewhat bearded
barbà-tus: barbed, bearded
barbíg-era: bearing barbs or beards
barbinér-vis: nerves bearded
barbinò-de: bearded at nodes
barbulà-tus: small-bearded
bartiseæfò-lius: bartisia-leaved
baselloì-des: basella-like

basilà-ris: pertaining to the base or bottom
bavár-icus: Bavarian
bellidifò-lius: beautiful-leaved
bellidioì-des: bellis-like
bél-lus: handsome
benedíc-tus: blessed
betà-ceus: beet-like
betonicæfò-lius, betonicifò-lius: betonica-leaved
betulæfò-lius: birch-leaved
betulì-nus: birch-like
betuloì-des: birch-like
bicarinà-tus: twice-keeled
bíc-olor: two-colored
bicór-nis, bicornù-tus: two-horned
bidentà-tus: two-toothed
bién-nis: biennial
bíf-idus: twice cut
biflò-rus: two-flowered
bifò-lius: two-leaved
bifór-mis: of two forms
bì-frons: two-fronded
bifurcà-tus: twice forked
bigíb-bus: with two swellings or projections
biglù-mis: two-glumed
bignonioì-des: bignonia-like
bíj-ugus: yoked, two together
bíl-obus: two-lobed
binà-tus: twin
binervà-tus, binér-vis: two-nerved
binoculà-ris: two-eyed, two-spotted
bipartì-tus: two-parted
bipét-alus: two-petaled
bipinnatíf-idus: twice pinnately cut
bipinnà-tus: twice pinnate
bipunctà-tus: two-spotted
biséc-tus: cut in two parts

143

biserrà-tus: twice toothed
bispinò-sus: two-spined
bistór-tus: twice twisted
bisulcà-tus: two-grooved
biternà-tus: twice ternate
bituminò-sus: bituminous, coal-black
bivál-vis: two-valved
blán-dus: bland, mild
blephariglót-tis: fringed-tongued
bò-nus: good
borbón-icus: of Bourbonne (France)
boreà-lis: northern
botryoì-des: cluster-like, grape-like
brachià-tus: branched at right angles
brachyán-drus: short-stamened
brachyán-thus: short-flowered
brachýb-otrys: short-clustered
brachycár-pus: short-fruited
brachypét-alus: short-petaled
brachýp-odus: short-stalked
brachýt-richus: short-haired
brachýt-ylus: short-styled, short-knobbed
bracteà-tus: bracteate, bearing bracts
bracteò-sus: bract-bearing
bractés-cens: bracteate
brasilià-nus: Brazilian
brassicæfò-lius: brassica-leaved
brevicaudà-tus: short-tailed
brevicaù-lis: short-stemmed
brevifò-lius: short-leaved
brév-ifrons: short-fronded
breviligulà-tus: short-liguled
brevipaniculà-tus: short-panicled
brevipedunculà-tus: short-peduncled
brév-ipes: short-footed or -stalked

brevirós-tris: short-beaked
brè-vis: short
breviscà-pus: short-scaped
brevisè-tus: short-bristled
brevís-pathus: short-spathed
brevís-simus: very short, short-est
brevís-tylus: short-styled
brilliantís-simus: very brilliant
brittán-icus: of Britain
brizæfór-mis: briza-formed
bronchià-lis: bronchial
brún-neus: deep brown
bucéph-alus: ox-headed
buddleifò-ilus: buddleja-leaved
buddleoì-des: buddleja-like
bufò-nius: pertaining to the toad
bulbíf-era: bulb-bearing
bulbò-sus: bulbous
bulgár-icus: Bulgarian
bullà-tus: blistered, puckered
bupleurifò-lius: bupleurum-leaved
buxifò-lius: box-leaved
byzantì-nus: Byzantine (Constantinople region)

cacaliæfò-lius: cacalia-leaved
cachemír-icus: of Cashmere
cád-micus: cadmic; metallic like tin
cærulés-cens: becoming dark blue
cærù-leus: cerulean, dark blue
cè-sius: bluish-gray
cæspitò-sus: cespitose, tufted
cáf-fer, cáf-fra: Kafir (Africa)
cajanifò-lius: cajanus-leaved (Cajan: pigeon-pea)
caláb-ricus: from Calabria (Italy)
calamifò-lius: reed-leaved
calathì-nus: basket-like

calcarà-tus: spurred
calcà-reus: pertaining to lime
calendulà-ceus: calendula-like
califór-nicus: of California
callicár-pus: beautiful-fruited
callistà-chyus: beautiful-spiked
callistegioì-des: callistegia-like
callizò-nus: beautiful-zoned
callò-sus: thick-skinned, with calluses
calocéph-alus: beautiful-headed
calóc-omus: beautiful-haired
calophýl-lus: beautiful-leaved
cál-vus: bald, hairless, naked
calýc-inus: calyx-like
calyculà-tus: calyx-like
calyptrà-tus: bearing a calyptra
cám-bricus: Cambrian, Welsh
campanulà-ria: bell-flowered
campanulà-tus: campanulate, bell-shaped
campanuloì-des: campanula-like
campés-tris: of the fields or plains
camphorà-tus: pertaining to camphor
campschát-icus: of Kamtchatka
campylocár-pus: curved-fruited
canaliculà-tus: channeled, grooved
cancellà-tus: cross-barred
candelà-brum: candelabra
cán-dicans: white, hoary
candidís-simus: very white-hairy or hoary
cán-didus: pure white, white-hairy, shining
canés-cens: gray-pubescent
canì-nus: pertaining to a dog
cannáb-inus: like cannabis or hemp
cantáb-ricus: from Cantabria (Spain)

cà-nus: ash-colored, hoary
capén-sis: of the Cape of Good Hope
capillà-ris: hair-like
capillifór-mis: hair-shaped
capíl-lipes: slender-footed
capità-tus: capitate, headed
capitella-tus: having little heads
capitél-lus: little head
capitulà-tus: having little heads
cappadóc-icum: Cappadocian (Asia Minor)
capreolà-tus: winding, twining
capricór-nis: Tropic of Capricorn
capsulà-ris: having capsules
cardaminefò-lius: cardamine-leaved
cardinà-lis: cardinal
cardiopét-alus: petals heart-shaped
carduà-ceus: thistle-like
caribǽ-us: of the Caribbean
caricò-sus: carex-like
carinà-tus: keeled
cariníf-era: keel-bearing
carminà-tus: carmine
cár-neus: flesh-colored
cár-nicus: fleshy
carniól-icus: of Carniola (south-central Europe)
carnós-ulus: somewhat fleshy
carnò-sus: fleshy
carolinià-nus, carolì-nus: Carolinian
carpáth-icus, carpát-icus: of the Carpathian region
carpinifò-lius: carpinus-leaved
cartilagín-eus: like cartilage
caryophyllà-ceus: clove-like
caryopteridifò-lius: caryopteris-leaved
caryotæfò-lius: caryota-leaved

caryotíd-eus: caryota-like
cashmerià-nus: of Cashmere (Asia)
cás-picus, cás-pius: Caspian
cassiaráb-icus: Arabian cassia
cassinoì-des: cassine-like
catalpifò-lius: catalpa-leaved
cathár-ticus: cathartic
cathayà-nus: of Cathay (China)
caucás-icus: belonging to the Caucasus
caudà-tus: caudate, tailed
caudés-cens: becoming stem-like
caulés-cens: having a stem
caulialà-tus: wing-stemmed
cauliflò-rus: stem-flowering
caús-ticus: caustic
celastrì-nus: celastrus-like
cenís-ius: of Mt. Cenis (France and Italy)
centifò-lius: hundred-leaved
centranthifò-lius: centranthus-leaved
cephalà-tus: bearing heads
cephalón-icus: of Cephalonia (one of the Ionian islands)
cephalò-tes: head-like
cerám-icus: ceramic, pottery-like
cerasíf-era: cerasus- or cherry-bearing
cerasifór-mis: cherry-formed
cerastioì-des: cerastium-like
ceratocaù-lis: horn-stalked
cereà-le: pertaining to Ceres or agriculture
cerefò-lius: wax-leaved
cè-reus: waxy
ceríf-era: wax-bearing
cerinthoì-des: cerinthe-like
cér-inus: waxy
cér-nuus: drooping, nodding
chalcedón-icus: of Chalcedon (on the Bosphorus)

chamædrifò-lius, chamædryfò-lius: chamædrys-leaved
chathám-icus: of Chatham Island (New Zealand)
cheilán-thus: lip-flowered
cheilanthifò-lius: cheilanthus-leaved
chelidonioì-des: chelidonium-like
chionán-thus: snow-flower
chirophýl-lus: hand-leaved
chloræfò-lius: chlora-leaved
chlorán-thus: green-flowered
chlorochì-lon: green-lipped
chrysanthemoì-des: chrysanthemum-like
chrysán-thus: golden-flowered
chrýs-eus: golden
chrysocár-pus: golden-fruited
chrysóc-omus: golden-haired
chrysól-epis: golden-scaled
chrysoleù-cus: gold and white
chrysól-obus: golden-lobed
chrysophýl-lus: golden-leaved
chrysós-tomus: golden-mouthed
chrysót-oxum: golden-arched
cichorià-ceus: cichorium-like
cicutæfò-lius: cicuta-leaved
cicutà-rius: of or like cicuta
cilià-ris, cilià-tus: ciliate, fringed
cilíc-icus: of Cilicia (Asia Minor)
ciliìc-alyx: calyx ciliate
ciliolà-ris: being secondarily ciliate
cínc-tus: girded, girdled
cinerariæfò-lius: cineraria-leaved
cinerás-cens: becoming ashy-gray
cinè-reus: ash-colored
cinnabarì-nus: cinnabar-red
cinnamò-meus: cinnamon-brown

cinnamomifò-lius: cinnamon-leaved

circinà-lis, circinà-tus: circinate, coiled

cirrhà-tus, cirrhò-sus: tendrilled

cismontà-nus: on this side the mountains

cisplatì-nus: on this side of La Plata River

cistifò-lius: cistus-leaved

citrà-tus: citrus-like

citrifò-lius: citrus-leaved

citrì-nus: citron-colored or -like

citriodò-rus: lemon-scented

citroì-des: citrus-like

cladóc-alyx: club-calyx

clandestì-nus: concealed

claù-sus: shut, closed

clavà-tus: club-shaped

clavellà-tus: slightly club-shaped

clà-vus: club

clematíd-eus: like clematis

clethroì-des: clethra-like

clivò-rum: of the hills

clypeà-tus: with, or like a shield

clypeolà-tus: somewhat shield-shaped

coarctà-tus: crowded together

coccíf-era, coccíg-era: berry-bearing

coccín-eus: scarlet

cochenillíf-era: cochineal-bearing

cochleà-ris: spoon-like

cochlearís-pathus: spoon-spathed

cochleà-tus: spoon-like

cœlestì-nus: sky-blue

cœlés-tis: celestial, sky-blue

cognà-tus: related to

cól-chicus: of Colchis (eastern Black Sea region)

collì-nus: pertaining to a hill

colorà-tus: colored

columbià-nus: Columbian (western North American)

columellà-ris: pertaining to a small pillar or pedestal

columnà-ris: columnar

cò-mans, comà-tus: with hair

commíx-tus: mixed, mingled

commù-nis: common, general

commutà-tus: changed or changing

comò-sus: with long hair

compác-tus: compact, dense

complanà-tus: flattened

compléx-us: circled, embraced

complicà-tus: complicate

compós-itus: compound

comprés-sus: compressed, flattened

cómp-tus: adorned, ornamented

cón-cavus: hollowed out

concín-nus: neat, well-made, elegant

conchæfò-lius: shell-leaved

cón-color: colored similarly

condensà-tus, condén-sus: condensed, crowded

confertiflò-rus: flowers crowded

confér-tus: crowded

confór-mis: similar in shape or otherwise

confù-sus: confused, uncertain

congés-tus: congested, brought together

conglomerà-tus: crowded together

congolà-nus: of the Congo

coníf-era: cone-bearing

conjugà-tus, conjugià-lis: connected, joined together

connà-tus: connate, united, twin

conoíd-eus: cone-like

conóp-seus: canopied

consanguín-eus: related
consól-idus: consolidated
conspér-sus: scattered
conspíc-uus: conspicuous
constríc-tus: constricted
contíg-uus: near together
continentà-lis: continental
contór-tus: contorted, twisted
contrác-tus: contracted
controvér-sus: controversial
convallarioì-des: convallaria-like
convolvulà-ceus: convolvulus-like
conyzoì-des: conyza-like
coralliflò-rus: coral-flowered
corál-linus: coral-red
cordà-tus: heart-shaped
cordifò-lius: heart-leaved
cordifór-mis: heart-form
corià-ceus: leathery
corià-ria: leather-like
coridifò-lius, corifò-lius, coriophýl-lus: coris-leaved
cór-neus: horny
corniculà-tus: horned
corníf-era, corníg-era: horn-bearing
cornù-tus: horned
corollà-tus: corolla-like
coromandelià-nus: of Coromandel (India)
coronà-rius: used for or belonging to garlands
coronà-tus: crowned
corrugà-tus: corrugated, wrinkled
cór-sicus: Corsican
corticò-sus: heavily furnished with bark
cortusoì-des: cortusa-like
corús-cans: vibrating, glittering
corylifò-lius: corylus-leaved

corymbíf-era: corymb-bearing
corymbiflò-rus: corymb-flowered
corymbò-sus: corymbose
corynóc-alyx: club-like calyx
cosmophýl-lus: cosmos-leaved
costà-tus: costate, ribbed
cotinifò-lius: cotinus-leaved (Cotinus, smoke-tree)
crassicaù-lis: thick-stemmed
crassifò-lius: thick-leaved
crás-sipes: thick-footed or -stalked
crassiús-culus: somewhat thick
crás-sus: thick, fleshy
cratægifò-lius: cratægus-leaved
crè-brus: close, frequent, repeated
crenatiflò-rus: crenate-flowered
crenà-tus: crenate, scalloped
crenulà-tus: crenulate, somewhat scalloped
crepidà-tus: slippered
crép-itans: crackling, rustling
cretà-ceus: pertaining to chalk
crét-icus: of Crete (island, E. Mediterranean)
crinì-tus: provided with long hair
crispà-tus, crís-pus: crisped, curled
Cristagál-li: cockscomb
cristà-tus: cristate, crested
crithmifò-lius: crithmum-leaved
crocà-tus: saffron-yellow
crò-ceus: saffron-colored, yellow
crocosmæflò-rus: crocosma-flowered
crotonifò-lius: croton-leaved
crucià-tus: cross-like
crucíf-era: cross-bearing
cruén-tus: bloody
Crusgál-li: cockspur
crustà-tus: encrusted

crystál-linus: crystalline
ctenoì-des: comb-like
cucullà-tus: hooded
cucumerì-nus: cucumber-like
cultò-rum: of the cultivators or gardeners
cultrà-tus: knife-shaped
cultrifór-mis: shaped like broad knife-blade
cuneà-tus: wedge-shaped
cuneifò-lius: wedge-leaved
cuneifór-mis: wedge-formed
cupreà-tus: coppery
cupressifór-mis: cypress-form
cuprés-sinus: cypress-like
cupressoì-des: cypress-like
cù-preus: copper-like or -colored
curassáv-icus: of Curaçoa (southern West Indies)
curvà-tus: curved
cúr-tus: shortened
curvifò-lius: leaves curved
cuscutæfór-mis: cuscuta-like
cuspidà-tus: with a cusp or sharp stiff point
cuspidifò-lius: leaves cuspidate
cyanán-thus: blue-flowered
cyà-neus: blue
cyanocár-pus: blue-fruited
cyanophýl-lus: blue-leaved
cyatheoì-des: cyathea-like
cyclamín-eus: cyclamen-like
cyclocár-pus: fruit rolled up circularly
cỳ-clops: cyclopean; gigantic
cylindrà-ceus, cylín-dricus: cylindrical
cylindrostà-chyus: cylindrical-spiked
cymbifór-mis: boat-shaped
cymò-sus: bearing cymes
cynán-chicus: cynanchum-like
cynanchoì-des: cynanchum-like

cynaroì-des: cynara-like
cỳ-preus: copper-like; see cupreus
cytisoì-des: cytisus-like

dacrydioì-des: dacrydium-like
dactylíf-era: finger-bearing
dactyloì-des: finger-like
dahù-ricus, daù-ricus, davù-ricus: of Dahuria or Dauria (Siberia)
dalmát-icus: Dalmatian
damascè-nus: of Damascus
daphnoì-des: daphne-like
dasyacán thus: thick-spined
dasyán-thus: thick-flowered
dasycár-pus: thick-fruited
dasýc-lados: thick-branched
dasyphýl-lus: thick-leaved
dasystè-mon: thick-stamened
daucoì-des: daucus-like
dealbà-tus: whitened, white-washed
déh-ilis: weak, frail
decán-drus: ten-stamened
decapét-alus: ten-petaled
decaphýl-lus: ten-leaved
decíd-uus: deciduous
decíp-iens: deceptive
declinà-tus: bent downward
decolò-rans: discoloring, staining
decompós-itus: decompound, more than once divided
déc-orans: adorning, decorative
decorà-tus: decorative
decò-rus: elegant, comely, becoming
decúm-bens: decumbent
decúr-rens: decurrent, running down the stem
defléx-us: bent abruptly downward

defór-mis: misshapen, **deformed**
dehís-cens: dehiscent
dejéc-tus: debased
deléc-tus: chosen
delicatís-simus: very delicate
delicà-tus: delicate, tender
deliciò-sus: delicious
delphinifò-lius: delphinium-leaved
deltoì-des, deltoíd-eus: triangular
demér-sus: under water
demís-sus: low, weak
dendroíd-eus: tree-like
densiflò-rus: densely flowered
densifò-lius: densely leaved
densà-tus: dense
dén-sus: dense
dentà-tus: toothed
denticulà-tus: slightly toothed
dentíf-era: tooth-bearing
dentò-sus: toothed
denudà-tus: denuded, naked
depauperà-tus: starved, dwarfed
depén-dens: hanging down
deprés-sus: depressed
desér-ti: of the desert
desmoncoì-des: desmoncus-like
detón-sus: clipped
deús-tus: burned
diaból-icus: diabolical, devilish
diacán-thus: two-spined
diadè-ma: diadem, crown
dián-drus: two-stamened
dianthiflò-rus: dianthus-flowered
diáph-anus: diaphanous, transparent
dichót-omus: forked in pairs
dichroán-thus: dichroa-flowered
   (Dichroa: Saxifragaceæ)
dích-rous: of two colors
dicóc-cus: with two berries

dictyophýl-lus: netted-leaved
díd-ymus: in pairs (as of stamens)
diffór-mis: of differing forms
diffù-sus: diffuse, spreading
digità-tus: digitate, hand-like
dilatà-tus: dilated, expanded
dilà-tus: dilated, spread out
dimidià-tus: halved
dimór-phus: two-formed
dì-odon: with two teeth
dioì-cus: diœcious
diosmæfò-lius: diosma-leaved
dipét-alus: two-petaled
diphýl-lus: two-leaved
diplostephioì-des: like diplostephium
dipsà-ceus: of the teasel or dipsacus
dipterocár-pus: two-winged fruit
díp-terus: two-winged
dipyrè-nus: two-seeded
discoíd-eus: discoid, rayless
dís-color: of two or of different colors
dís-par: dissimilar, unlike
disséc-tus: dissected, deeply cut
dissím-ilis: unlike
dissitiflò-rus: remotely or loosely flowered
distà-chyus: two-spiked
dís-tans: distant, separate, remote
distichophýl-lus: leaves two-ranked
dís-tichus: two-ranked
dís-tylus: two-styled
diúr-nus: day-flowering
divaricà-tus: spreading, widely divergent
divér-gens: wide-spreading
diversíc-olor: diversely colored

diversiflò-rus: diversely flowered
diversifò-lius: variable-leaved
divì-sus: divided
dixán-thus: double-tinted
dodecán-drus: twelve-stamened
dodonæifò-lius: dodonæa-leaved
dolabrifór-mis: hatchet-shaped
dolabrà-tus: mattock- or hatchet-shaped
dolò-sus: deceitful
domés-ticus: domestic, domesticated
doronicoì-des: doronicum-like
drabifò-lius: draba-leaved
dracænoì-des: dracæna-like
dracocéph-alus: dragon-head
dracunculoì-des: tarragon-like
drepanophýl-lus: leaves sickle-shaped
drupà-ceus: drupe-like
drupíf-era: drupe-bearing
drynarioì-des: drynaria-like
dù-bius: doubtful
dúl-cis: sweet
dumetò-rum: of bushes or hedges
dumò-sus: bushy
dù-plex: double
duplicà-tus: duplicate, double
duráb-ilis: durable, lasting
durác-inus: hard-berried
dù-rior: harder
duriús-culus: somewhat hard or rough

ebenà-ceus: ebony-like
ebracteà-tus: bractless
ebúr-neus: ivory-white
echinà-tus: prickly, bristly
echinocár-pus: prickly-fruited
echinosép-alus: prickly-sepaled
echioì-des: echium-like
ecornù-tus: hornless
edù-lis: edible

effù-sus: very loose-spreading
elæagnifò-lius: elæagnus-leaved
elás-ticus: elastic
elà-tior, elà-tius: taller
elà-tus: tall
él-egans: elegant
elegantís-simus: most elegant
elegán-tulus: elegant
elephán-tidens: large-toothed
elephán-tipes: elephant-footed
elephán-tum: of the elephants
ellipsoidà-lis: elliptic
ellíp-ticus: elliptic
elongà-tus: elongated, lengthened
emarginà-tus: with a shallow notch at apex
emét-icus: emetic
ém-inens: eminent, prominent
empetrifò-lius: empetrum-leaved
enneacán-thus: nine-spined
enneaphýl-lus: nine-leaved
ensà-tus: sword-shaped
ensifò-lius: sword-leaved
ensifór-mis: sword-shaped
entomóph-ilus: insect-loving
equés-tris: pertaining to the horse
equisetifò-lius: equisetum-leaved
equì-nus: of horses
eréc-tus: erect, upright
eriacán-thus: woolly-spined
erianthè-ra: woolly-anthered
erián-thus: woolly-flowered
ericæfò-lius, ericifò-lius: erica-leaved
ericoì-des: erica-like, heath-like
erinà-ceus: hedge-hog
eriobotryoì-des: eriobotrya-like
eriocár-pus: woolly-fruited
eriocéph-alus: woolly-headed
erióph-orus: wool-bearing

151

eriós-pathus: woolly-spathed
eriostà-chyus: woolly-spiked
eriostè-mon: stamens woolly
erò-sus: erose, jagged, as if gnawed
errát-icus: erratic, unusual, sporadic
erubés-cens: blushing
erucoì-des: eruca-like
erythrocár-pus: red-fruited
erythrocéph-alus: red-headed
erythróp-odus: red-footed, red-stalked
erythróp-terus: red-winged
erythrosò-rus: red sori
esculén-tus: esculent, edible
estrià-tus: without stripes
etrús-cus: Etruscan (in Italy)
etuberò-sus: without tubers
eucalyptoì-des: eucalyptus-like
eugenioì-des: eugenia-like
eupatorioì-des: eupatorium-like
euphorbioì-des: euphorbia-like
europǽ-us: European
evéc-tus: extended
evér-tus: expelled, turned out
exaltà-tus: exalted, very tall
exarà-tus: furrowed
excavà-tus: excavated, hollowed out
excél-lens: excellent, excelling
excél-sus: tall
excél-sior: taller
excì-sus: cut away
excorticà-tus: stripped of bark
exíg-uus: little, small, poor
exím-ius: distinguished, out of the ordinary
exitiò-sus: pernicious, destructive
exolè-tus: mature, dying away
exót-icus: exotic, from another country
expán-sus: expanded

explò-dens: exploding
exscà-pus: without scape
exscúlp-tus: dug out
exsér-tus: protruding from
exsúr-gens: rising up
extén-sus: extended
exù-dans: exuding

fabà-ceus: faba-like, bean-like
falcà-tus: falcate, sickle-shaped
falcifò-lius: falcate-leaved
falcifór-mis: sickle-shaped
fál-lax: deceptive
farinà-ceus: containing starch
fariníf-era: starch-bearing
farinò-sus: mealy, powdery
fascià-tus: abnormally flattened
fasciculà-ris. fasciculà-tus: fascicled, clustered
fascinà-tor: fascinating
fastigià-tus: fastigiate, branches erect and close together
fastuò-sus: proud
fát-uus: foolish, simple
febríf-ugus: fever-dispelling
fém-ina: female
fenestrà-lis: with window-like openings
fè-rox: ferocious, very thorny
fér-reus: pertaining to iron
ferrugín-eus: rusty
fér-tilis: fertile, fruitful
ferulæfò-lius: ferula-leaved
festì-vus: festive, gay, bright
fibrillò-sus: having fibers
fibrò-sus: having prominent fibers
ficifò-lius: fig-leaved
ficoì-des, ficoíd-eus: fig-like
filamentò-sus: filamentous
filicà-tus: fern-like
filicaù-lis: thread-stemmed
filicifò-lius: fern-leaved

filicì-nus: fern-like
filicoì-des: fern-like
filíf-era: bearing filaments or threads
filifò-lius: thread-leaved
filifór-mis: thread-like
filipendulì-nus: filipendula-like
fíl-ipes: stalks thread-like
fimbriát-ulus: with small fringe
fimbrià-tus: fringed
firmà-tus: firm, made firm
fír-mus: firm, strong
fissifò-lius: split-leaved
fís-silis: cleft or split
fissurà-tus: fissured, cleft
fís-sus: cleft, split
fistulò-sus: hollow-cylindrical
flabellà-tus: with fan-like parts
flabél-lifer, flabellifór-mis: fan-shaped
flác-cidus: flaccid, soft
flagellà-ris, flagellà-tus: whip-like
flagellifór-mis: whip-formed
flagél-lum: a scourge or flail
flám-meus: flame-colored
flavés-cens: yellowish
flavíc-omus: yellow-wooled or -haired
fláv-idus: yellow, yellowish
flavispì-nus: yellow-spined
flavís-simus: deep yellow, very yellow
flà-vus: yellow
flexicaù-lis: pliant-stemmed
fléx-ilis: flexible, pliant, limber
flexuò-sus: flexuose, tortuous, zig-zag
floccò-sus: woolly
flò-re-ál-bo: with white flowers
florentì-nus: Florentine
flò-re-plè-no: with double flowers

floribún-dus: free-flowering
floridà-nus: of Florida
flór-idus: flowering, full of flowers
flù-itans: floating
fluviát-ilis: pertaining to a river
fœm-ina: feminine
fœniculà-tus: fennel-like
fœtidís-simus: very fetid
fœt-idus: fetid, bad-smelling
folià-ceus: leaf-like
folià-tus: with leaves
foliolà-tus: with leaflets
foliolò-sus: having leaflets
foliò-sus: leafy, full of leaves
folliculà-ris: bearing follicles
fontinà-lis: pertaining to a spring of water
forficà-tus: shear-shaped
fornicæfór-mis: ant-shaped
formosà-nus: of Formosa
formosís-simus: very beautiful
formò-sus: beautiful, handsome
fourcroỳ des: like fourcroya
foveà-tus: pitted
foveolà-tus: pitted
fragarioì-des: strawberry-like
frág-ilis: fragile, brittle
frà-grans: fragrant
fragrantís-simus: very fragrant
fraxín-eus: like fraxinus
fraxinifò-lius: fraxinus-leaved
fríg-idus: cold, of cold regions
frondò-sus: leafy
fructíf-era: fruit-bearing, fruit-ful
fructíg-enus: fruitful
frumentà-ceus: pertaining to grain
frutés-cens: shrubby, bushy
frù-tex: a shrub or bush
frù-ticans: shrubby, shrub-like
fruticò-sus: shrubby, bushy

fucà-tus: painted, dyed
fuchsioì-des: fuchsia-like
fù-gax: swift
fúl-gens: shining, glistening
fúlg-idus: fulgid, shining
fuliginò-sus: sooty, black-colored
fulvés-cens: fulvous
fúl-vidus: slightly tawny
fúl-vus: fulvous, tawny, orange-gray-yellow
fumariæfò-lius: fumaria-leaved
fù-nebris: funereal
fungò-sus: fungous, spongy
funiculà-tus: of a slender rope or cord
fúr-cans, furcà-tus: furcate, forked
furfurà-ceus: scurfy
fuscifo-lius: fuscous-leaved
fús-cus: fuscous, brown, dusky
fusifór-mis: spindle-shaped

galacifò-lius: galax-leaved
galán-thus: milk flower
galeà-tus: helmeted
galegifò-lius galega-leaved
galericulà-tus: helmet-like
galioì-des: galium-like
gál-licus: of Gaul or France; also pertaining to a cock or rooster
gangét-icus: of the Ganges
gargán-icus: belonging to Gargano (Italy)
gél-idus: ice-cold
geminà-tus: twin
geminiflò-rus: twin-flowered
geminispì-nus: twin-spined
gemmà-tus: bearing buds
gemmíf-era: bud-bearing
generà-lis: general, prevailing
geniculà-tus: jointed, kneed

genistifò-lius: genista-leaved
geoì-des: of the earth
geomét-ricus: in a pattern
geonomæfór-mis: geonoma-formed
georgià-nus: of Georgia
geranioì-des: geranium-like
germán-icus: German
gibberò-sus: humped, hunchbacked
gibbiflò-rus: gibbous-flowered
gibbò-sus, gíb-bus: swollen on one side
gibraltár-icus: of Gibraltar
gigantè-us: gigantic, very large
gigán-thes: giant-flowered
gì-gas: of giants, immense
glabél-lus: smoothish
glà-ber: glabrous, smooth
glabér-rimus: very smooth, smoothest
glabrà-tus: somewhat glabrous
glabrés-cens: smoothish
glacià-lis: icy, frozen
gladià-tus: sword-like
glandifór-mis: gland-formed
glandulíf-era: gland-bearing
glandulò-sus: glandular
glaucés-cens: becoming glaucous
glaucifò-lius: glaucous-leaved
glaucoì-des: glaucous-like
glaucophýl-lus: glaucous-leaved
glaù-cus: glaucous, with a bloom
globò-sus: globose, spherical
globulà-ris: of a little ball or sphere
globulíf-era: globule- or globe-bearing
globulò-sus: like a little ball
glomerà-tus: glomerate, clustered
glomeruliflò-rus: flowers in glomerules

gloriò-sus: glorious, superb
gloxinioì-des: gloxinia-like
glumà-ceus: with glumes or glume-like structures
glutinò-sus: glutinous, sticky
glycinioì-des: glycine-like
gnaphalò-des: gnaphalium-like (Gnaphalium, a Composite)
gomphocéph-alus: club-headed
gomphocóc-cus: club-berry
gongylò-des: roundish, swollen
gonià-tus: angled, cornered
goniòc-alyx: calyx cornered
gossýp-inus: gossypium-like, cotton-like
gracilén-tus: slender
graciliflò-rus: graceful-flowered
gracíl-ior: more graceful
gracíl-ipes: slender-footed
gràc-ilis: graceful, slender
gracilís-tylus: slender-styled
gracíl-limus: very slender
græ̀-cus: Greek, of Greece
gramín-eus: grassy, grass-like
graminifò-lius: grass-leaved
grammopét-alus: petals striped or marked
gràn-diceps: large-headed
grandicús-pis: with large cusps or points
grandidentà-tus: large-toothed
grandiflò-rus: large-flowered
grandifò-lius: large-leaved
grandifór-mis: on a large scale
grandipunctà-tus: with large spots
gràn-dis: large, big
granít-icus: granite-loving
granulà-tus: granulate, covered with minute grains
granulò-sus: granulate
gratís-simus: very pleasing or agreeable

grà-tus: pleasing, agreeable
gravè-olens: heavy-scented
grís-eus: gray
grœnlán-dicus: of Greenland
gròsse-serrà-tus: large-toothed
grù-inus: of a crane
gummíf-era: gum-bearing
gunneræfò-lius: gunnera-leaved
guttà-tus: spotted, speckled
gymnocár-pus: naked-fruited
gymnocaù-lon: slender-stemmed
gymnocéph-alus: slender-headed
gỳ-rans: revolving in a circle, gyrating

hadriát-icus: Adriatic
hæmán-thus: blood-red-flowered
hæmastò-mus: red-mouthed
hæmatóc-alyx: calyx blood-red
hæmatò-des: bloody
hakeoì-des: hakea-like
halimifò-lius: halimium-leaved
halóph-ilus: salt-loving
hamà-tus, hamò-sus: hooked
harpophýl-lus: sickle-leaved
hastà-tus: hastate, spear-shaped
hastíf-era: spear-bearing
hastilà-bium: halbert-lipped
hastì-lis: of a javelin or spear
hastulà-tus: somewhat spear-shaped
hebecár-pus: pubescent-fruited
hebephýl-lus: pubescent-leaved
hederà-ceus: of the ivy
helianthoì-des: helianthus-like
helvét-icus: Swiss
hél-volus: pale yellow
hemiphlœ̀-us: half-barked
hemisphær-icus: hemispherical
hepaticæfò-lius: hepatica-leaved
heptaphýl-lus: seven-leaved
heracleifò-lius: heracleum-leaved

herbà-ceus: herbaceous, not woody

hespér-ius: of the West

heteracán-thus: various-spined

heterán-thus: various-flowered

heterocár-pus: various-fruited

hetér-odon: various-toothed

heterodóx-us: heterodox

heteroglós-sus: various-tongued

heteról-epis: variable-scaled

heteromór-phus: various in form

heteropét-alus: various-petaled

heterophýl-lus: various-leaved

heteróp-odus: various-footed or -stalked

hexagonóp-terus: six-angled-winged

hexagò-nus: six-angled

hexán-drus: with six stamens

hexapét-alus: six-petaled

hexaphýl-lus: six-leaved

hì-ans: open, gaping

hibernà-lis: pertaining to winter

hibér-nicus: of Ireland

hibiscifò-lius: hibiscus-leaved

hierochún-ticus: of Jericho

hieroglýph-icus: marked as if with signs

himalà-icus: Himalayan

hircì-nus: with a goat's odor

hirsutís-simus: very hairy

hirsù-tulus: somewhat hairy

hirsù-tus: hirsute, hairy

hirtél-lus: somewhat hairy

hirtiflò-rus: hairy-flowered

hír-tipes: hairy-stalked or -stemmed

hír-tus: hairy

hispán-icus: Spanish

hispidís-simus: very bristly

hispíd-ulus: somewhat bristly

his-pidus: hispid, bristly

hollán-dicus: of Holland

holocár-pus: whole-fruited

holochrỳ-sus: wholly golden

holoseríc-eus: woolly-silky

homól-epis: homologous scales

horizontà-lis: horizontal

hór-ridus: prickly, horridly armed

hortén-sis, hortò-rum, hortulà-nus, hortulà-lis, hortulò-rum: belonging to a hortus or garden, or to gardens

humifù-sus: sprawling on the ground

hù-milis: low-growing, dwarf

humilifò-lius: hop-leaved

hyacínth-inus: sapphire-colored

hyacinthoì-des: hyacinth-like

hyál-inus: transparent, translucent

hýb-ridus: hybrid, mixed, mongrel

hydrangeoì-des: hydrangea-like

hyemà-lis: of winter

hygromét-ricus: taking up water

hymenán-thus: membranaceous-flowered

hymenò-des: membrane-like

hymenorrhì-zus: membranous-rooted

hymenosép-alus: sepals membranous

hyperbò-reus: far northern

hypericifò-lius: hypericum-leaved

hypericoì-des: hypericum-like

hypnoì-des: moss-like

hypocraterifór-mis: salver-shaped

hypogà̀-us: underground

hypoglaù-cus: glaucous beneath

hypoglót-tis: under-tongued

hypoleù-cus: whitish, pale beneath

hypophýl-lus: under the leaf
hyrcà-nium: Hyrcanian (near Caspian Sea)
hyssopifò-lius: hyssop-leaved
hýs-trix: porcupine-like, bristly
ián-thinus: violet, violet-blue
ibér-icus, iberíd-eus: of Iberia (Spain, Portugal)
iheridifò-lius: iberis-leaved
icosán-drus: twenty-stamened
idæ̀-us: of Mt. Ida (Asia Minor)
ignés-cens: fiery
ig-neus: fiery
ilicifò-lius: ilex-leaved, holly-leaved
illecebrò-sus: of the shade
illinì-tus: varnished
illustrà-tus: pictured
illús-tris: bright, brilliant, lustrous
illýr-icus: of Illyria (ancient region of southern Europe)
imberbiflò-rus: flowers beardless
imbér-bis: without beards or spines
ím-bricans: imbricating
imbrìcà-tus: imbricated, lapping over
immaculà-tus: immaculate, spotless
immér-sus: under water
impà-tiens: impatient
imperà-tor: commanding, imperious
imperià-lis: imperial, kingly
impléx-us: interwoven
imprés-sus: impressed, sunken in
inæqualifò-lius: unequal-leaved
inæquà-lis: unequal
inæquilát-erus: unequal-sided
incà-nus: hoary
incarnà-tus: flesh-colored

incér-tus: uncertain, doubtful
incisifò-lius: cut-leaved
incì-sus: incised, cut
inclaù-dens: never-closing
inclinà-tus: bent downward
incomparáb-ilis: incomparable, excelling
incómp-tus: rude, unadorned
inconspíc-uus: inconspicuous
incrassà-tus: thickened
incurvà-tus, incúr-vus: incurved, bent inward
indentà-tus: indented
in-dicus: of India
indivì-sus: undivided
inér-mis: unarmed
infaù-stus: unfortunate
infectò-rius: pertaining to dyes
infés-tus: dangerous, unsafe
inflà-tus: inflated, swollen up
infortunà-tus: unfortunate
infrác-tus: broken
infundibulifór-mis: funnelform, trumpet-shaped
infundíb-ulum: a funnel
ín-gens: enormous
inodò-rus: without odor
inornà-tus: without ornament
ín-quinans: polluting, discoloring
inscríp-tus: written on
insíg-nis: remarkable, distinguished, marked
insitít-ius: grafted
insulà-ris: insular
intác-tus: intact, untouched
ín-teger: entire
integér-rimus: very entire
integrifò-lius: entire-leaved
interjéc-tus: interjected, put between
intermè-dius: intermediate
interrúp-tus: interrupted

intertéx-tus: interwoven, intertwined

intór-tus: twisted

intricà-tus: intricate, entangled

intrór-sus: introrse, turned inward

intumés-cens: swollen, puffed up, tumid

intybà-ceus: pertaining to chicory

invér-sus: inverse, turned over

invì-sus: unseen, overlooked

involucrà-tus: with an involucre

involù-tus: rolled inward

ionán-drus: violet-anthered

ionán-thus: violet-flowered

ionóp-terus: violet-winged

iridés-cens: iridescent

iridiflò-rus: iris-flowered

irregulà-ris: irregular

irríg-uus: watered

isán-drus: with equal stamens

isopét-alus: equal-petaled

isophýl-lus: equal-leaved

ís-tria: of Istria (southern Europe)

itál-icus: Italian

ixioì-des: ixia-like

ixocár-pus: sticky- or glutinous-fruited

japón-icus: Japanese

jasmín-eus: jasmine-like

jasminiflò-rus: jasmine-flowered

jasminoì-des: jasmine-like

javán-icus: of Java

jubà-tus: crested, with a mane

jucún-dus: agreeable, pleasing

jugò-sus: joined, yoked

jún-ceus: juncus-like, rush-like

juncifò-lius: rush-leaved

juniperifò-lius: juniper-leaved

juniperì-nus: juniper-like; sometimes bluish-brown, like berries of juniper

kamtschát-icus: of Kamtchatka

kashmirià-nus: of Cashmere

koreà-nus, korià-nus, koraién-sis: of Korea

labià-tus: labiate, lipped

láb-ilis: slippery

labiò-sus: lipped

labrò-sus: large-lipped

laburnifò-lius: laburnum-leaved

lác-erus: torn

lacinià-tus: laciniate, torn

laciniò-sus: much laciniate

lactà-tus: milky

lác-teus: milk-white

lactíc-olor: milk-colored

lactíf-era: milk-bearing

lactiflò-rus: flowers milk-colored

lacunò-sus: with holes or pits

lacús-tris: pertaining to lakes

ladaníf-era, ladán-ifer: ladanum-bearing (resinous juice)

lætiflò-rus: bright- or pleasing-flowered

lætév-irens: light or vivid green

lǽ-tus: bright, vivid

lævicaù-lis: smooth-stemmed

lævigà-tus: smooth

lǽv-ipes: smooth-footed

lǽ-vis: smooth

læviús-culus: smoothish

lagenà-rius: of a bottle or flask

lanà-tus: woolly

lanceifò-lius: lance-leaved

lanceolà-tus: lanceolate

lán-ceus: lance-like

lancifò-lius: lance-leaved

laníg-era: wool-bearing

lán-ipes: woolly-footed or -stalked
lanò-sus: woolly
lanuginò-sus: woolly, downy
lappà-ceus: lappa-like
lappón-icus: of Lapland
laricifò-lius: larch-leaved
laríc-inus: larch-like
lasiacán-thus: pubescent-spined
lasián-drus: pubescent-stamened
lasián-thus: woolly-flowered
lasiocár-pus: rough- or woolly-fruited
lasiodón-tus: woolly-toothed
lasioglós-sus: tongue rough-hairy
lasiól-epis: woolly-scaled
lasiopét-alus: petals rough-hairy
lateriflò-rus: lateral-flowered
latér-ipes: lateral-stalked
laterít-ius: brick-red
latiflò-rus: broad-flowered
latifò-lius: broad-leaved
lát-ifrons: broad-fronded
latilà-brus: broad-lipped
latíl-obus: broad-lobed
latimaculà-tus: broad-spotted
lát-ipes: broad-footed or -stalked
latispì-nus: broad-spined
latisquà-mus: broad-scaled
latís-simus: broadest, very broad
là-tus: broad, wide
laudà-tus: lauded, worthy
laurifò-lius: laurel-leaved
laurì-nus: laurel-like
lavandulà-ceus: lavender-like
lavateroì-des: lavatera-like
laxiflò-rus: loose-flowered
laxifò-lius: loose-leaved
láx-us: lax, open, loose
ledifò-lius: ledum-leaved
   (Ledum: Ericaceæ)
leián-thus: smooth-flowered

leiocár-pus: smooth-fruited
leióg-ynus: smooth pistil
leiophýl-lus: smooth-leaved
lenticulà-ris, lentifór-mis: lenticular, lens-shaped
lentiginò-sus: freckled
lentiscifò-lius: lentiscus-leaved
   (Lentiscus: Pistacia)
lén-tus: pliant, tenacious, tough
leontoglós-sus: lion-tongued or -throated
leopardì-nus: leopard-spotted
lepidophýl-lus: scaly-leaved
lepidò-tus: with small scurfy scales
lép-idus: graceful, elegant
leprò-sus: scurfy
leptán-thus: thin-flowered
leptocaù-lis: thin-stemmed
leptóc-ladus: thin-stemmed or -branched
leptól-epis: thin-scaled
leptopét-alus: thin-petaled
leptophýl-lus: thin-leaved
leptosép-alus: thin-sepaled
lép-topus: thin- or slender-stalked
leptostà-chyus: thin-spiked
lepturoì-des: lepturus-like (Lepturus: Gramineæ)
leucanthemifò-lius: leucanthemum-leaved
leucán-thus: white-flowered
leucób-otrys: with white clusters
leucocaù-lis: white-stemmed
leucocéph-alus: white-headed
leucochì-lus: white-lipped
leucodér-mis: white-skinned
leuconeù-rus: white-nerved
leucophæ̀-us: dusky-white
leucophýl-lus: white-leaved
leucorhì-zus: white-rooted
leucós-tachys: white-spiked

leucót-riche: white-haired
leucoxán-thus: whitish-yellow
leucóx-ylon: white-wooded
libanót-icus: of Libania
libúr-nicus: of Liburnia
lignò-sus: woody
ligulà-ris, ligulà-tus: ligulate, strap-shaped
ligús-ticus: of Liguria
ligusticifò-lius: ligusticum-leaved (Ligusticum: Umbel-liferæ)
ligustrifò-lius: privet-leaved
ligustrì-nus: privet-like
lilác-inus: lilac
lilià-ceus: lily-like
liliiflò-rus: lily-flowered
lilifò-lius: lily-leaved
limbà-tus: bordered
limonifò-lius: lemon-leaved
limò-sus: of muddy or marshy places
linariifò-lius: linaria-leaved
linarioì-des: linaria-like
linearifò-lius: linear-leaved
linearìl-obus: linear-lobed
lineà-ris: linear
lineà-tus: lined, with lines or stripes
linguefór-mis: tongue-shaped
lingulà-tus: tongue-shaped
liniflò-rus: flax-flowered
linifò-lius: flax-leaved
linnæoì-des: linnæa-like
linoì-des: flax-like
linophýl-lus: flax-leaxed
lithóph-ilus: dwelling on rocks
lithospér-mus: seeds stone-like
littorà-lis: of the seashore
lituiflò-rus: trumpet-flowered
lív-idus: livid, bluish
lobà-tus: lobed
lobelioì-des: lobelia-like

lobocár-pus: lobed-fruited
lobophýl-lus: lobed-leaved
lobulà-ris: lobed
lobulà-tus: with small lobes
lolià-ceus: lolium-like
longebracteà-tus: long-bracted
longepedunculà-tus: long-peduncled
longicaudà-tus: long-tailed
longicaù-lis: long-stemmed
longíc-omus: long-haired
longicús-pis: long-pointed
longiflò-rus: long-flowered
longifò-lius: long-leaved
longihamà-tus: long-hooked
longíl-abris: long-lipped
longilaminà-tus: with long plates
longíl-obus: long-lobed
longimucronà-tus: long-mucro-nate
lóng-ipes: long-footed or -stalked
longipét-alus: long-petaled
longipinnà-tus: long-pinnate
longiracemò-sus: long-racemed
longirós-tris: long-beaked
longiscà-pus: long-scaped
longisép-alus: long-sepaled
longís-pathus: long-spathed
longispì-nus: long-spined
longís-simus: longest, very long
longís-tylus: long-styled
lón-gus: long
lophán-thus: crest-flowered
lorifò-lius: strap-leaved
lotifò-lius: lotus-leaved
lousià-nus: of Louisiana
lù-cidus: lucid, bright, shining, clear
ludovicià-nus: of Louisiana
lunà-tus: lunate, crescent-shaped
lunulà-tus: somewhat crescent-shaped
lupulì-nus: hop-like

160

lù-ridus: lurid, wan, pale yellow

lusitán-icus: of Portugal

lutè-olus: yellowish

lutés-cens: becoming yellowish

lutetià-nus: Parisian

lù-teus: yellow

luxù-rians: luxuriant, thrifty

lychnidifò-lius: lychnis-leaved

lycóc-tonum: wolf-poison

lycopodiòi-des: lycopodium-like, clubmoss-like

lyrà-tus: lyrate, pinnatifid with large terminal lobe

lysimachiòi-des: lysimachia-like

macedón-icus: Macedonian

macilén-tus: lean, meager

macracán-thus: large-spined

macrán-drus: with large anthers

macrán-thus: large-flowered

macradè nia, macrodè-num: large-glanded

macro-: long, but often large or big. See page 126

maculà-tus, maculò-sus: spotted

mæsì-acus: of Mœsia, ancient name of Bulgaria and Serbia

magellán-icus: Straits of Magellan region

magníf-icus: magnificent, distinguished

mág-nus: large

majà-lis: of May, Maytime

majés-ticus: majestic

mà-jor, mà-jus: greater, larger

malabár-icus: of Malabar

malacòi-des: soft, mucilaginous

malacospér-mus: soft-seeded

malifór-mis: apple-formed

malvà-ceus: mallow-like

malvæflò-rus: mallow-flowered

mamillà-tus, mammillà-ris, mammò-sus: with breasts or nipples

mammulò-sus: with small nipples

mandshù-ricus, mandschú-ricus: of Manchuria

manicà-tus: manicate, long-sleeved

manzanì-ta: little apple

margarità-ceus: pearly, of pearls

margaritíf-era: pearl-bearing

marginà-lis: marginal

marginà-tus· margined

marginél-lus: somewhat margined

marià-nus: of Maryland

marilán-dicus, marylán-dicus: of Maryland

marít-imus: maritime, of the sea

marmorà-tus, marmò-reus: marbled, mottled

marmophýl-lus: leaves marbled

maroccà-nus: of Morocco

más, masculà-tus, más-culus: male, masculine

matricariæfò-lius: matricaria-leaved

matronà-lis: pertaining to matrons

mauritán-icus: of Mauretania (northern Africa)

maxillà-ris: maxillary, of the jaw

máx-imus: largest

méd-icus: medicinal

mediopíc-tus: pictured or striped at the center

mediterrà-neus: of the Mediterranean region

mè-dius: medium, intermediate

medullà-ris: of the marrow or pith

megacán-thus: large-spined
megacár-pus: large-fruited
megalán-thus: large-flowered
megalophýl-lus, megaphýl-lus: large-leaved
megapotám-icus: of the big river
megarrhì-zus: large-rooted
megaspér-mus: large-seeded
megastà-chyus: large-spiked
megastíg-mus: with large stigmas
meiacán-thus: small-flowered
melanán-thus: black-flowered
melanocén-trus: black-centered
melanchól-icus: melancholy, hanging or drooping
melanocár-pus: black-fruited
melanocaù-lon: black-stemmed
melanocóc-cus: black-berried
melanoleù-cus: black and white
melanóx-ylon: black-wooded
melanthè-rus: black-anthered
meleà-gris: like a guinea-fowl, speckled
mél-leus: pertaining to honey
mellíf-era: honey-bearing
melliodò-rus: honey-scented
mellì-tus: honey-sweet
melofór-mis: melon-shaped
membranà-ceus: membranaceous
meniscifò-lius: crescent-leaved
meridionà-lis: southern
mesoleù-cus: mixed with white
metál-licus: metallic
meteloì-des: metel-like
mexicà-nus: Mexican
mì-cans: glittering, sparkling
michauxioì-des: michauxia-like (Michauxia of Campanulaceæ)
micracán-thus: small-spined
micrán-thus: small-flowered
microcár-pus: small-fruited

microcéph-alus: small-headed
microchì-lum: small-lipped
micród-asys: small, thick, shaggy
míc-rodon: small-toothed
microglós-sus: small-tongued
micról-epis: small-scaled
micróm-eris: small number of parts
micropét-alus: small-petaled
microphýl-lus: small-leaved
micróp-terus: small-winged
microsép-alus: small-sepaled
microstè-mus: of small filaments
microthè-le: small nipple
mikanioì-des: mikania-like (Mikania of Compositæ)
milià-ceus: pertaining to millet
militò-ris: military
millefolià-tus, millefò-lius: thousand-leaved
mimosoì-des: mimosa-like
mì-mus: mimic
mì-nax: threatening, forbidding
minià-tus: cinnabar-red
mín-imus: least, smallest
mì-nor, mì-nus: smaller
minutiflò-rus: minute-flowered
minutifò-lius: minute-leaved
minutís-simus: very or most minute
minù-tus: minute, very small
miráb-ilis: marvellous, extraordinary
mì-tis: mild, gentle
mitrà-tus: turbaned
míx-tus: mixed
modés-tus: modest
mœsì-acus: of the Balkan region
moldáv-icus: of Moldavia (Danube region)
mól-lis: soft, soft-hairy
mollís-simus: very soft-hairy
moluccà-nus: of the Moluccas

(East Indies)
monacán-thus: one-spined
monadél-phus: in one group or bundle
monán-drus: one-stamened
mongól-icus: of Mongolia
monilíf-era: bearing a necklace
monocéph-alus: single-headed
monóg-ynus: of one pistil
monoì-cus: monœcious
monopét-alus: one-petaled
monophýl-lus: one-leaved
monóp-terus: one-winged
monopyrè-nus: bearing one stone or pyrene
monosép-alus: one-sepaled
monospér-mus: one-seeded
monostà-chyus: one-spiked
monspessulà-nus: of Montpelier
monstrò-sus: monstrous, abnormal
montà-nus: pertaining to mountains
montén-sis: citizen of mountains
montíc-olus: inhabiting mountains
montíg-enus: mountain-born
morifò-lius: morus-leaved; mulberry-leaved
mosà-icus: parti-colored
moschà-tus: musky
mucò-sus: slimy
mucronà-tus: mucronate
mucronulà-tus: with a small mucro or point
multibracteà-tus: many-bracted
multicaù-lis: many-stemmed
multíc-avus: with many hollows
múl-ticeps: many-headed
multíc-olor: many-colored
multicostà-tus: many-ribbed
multíf-idus: many times parted

multiflò-rus: many-flowered
multifurcà-tus: much-forked
multíj-ugus: many in a yoke
multilineà-tus: many-lined
multinér-vis: many-nerved
múl-tiplex: many-folded
multiradià-tus: with numerous rays
multiséc-tus: much cut
mún-dulus: trim, neat
munì-tus: armed, fortified
murà-lis: of walls
muricà-tus: muricate, roughed by means of hard points
musà-icus: musa-like
muscætóx-icum: fly-poison
muscíp-ula: fly-catcher
muscoì-des: moss-like
muscív-orus: fly-eating
muscò-sus: mossy
mutáb-ilis, mutà-tus: changeable, variable
mù-ticus: blunt, pointless
mutilà-tus: mutilated
myoporoì-des: myoporum-like
myriacán-thus: myriad-spined
myriocár-pus: myriad-fruited
myrióc-ladus: myriad-branched
myriophýl-lus: myriad-leaved
myriostíg-mus: myriad-stigmaed
myrmecóph-ilus: ant-loving
myrsinifò-lius: myrsine-leaved
myrsinoì-des: myrsine-like
myrtifò-lius: myrtle-leaved

nanél-lus: very dwarf
nà-nus: dwarf
napifór-mis: turnip-shaped
narcissiflò-rus: narcissus-flowered
narinò-sus: broad-nosed
nasù-tus: large-nosed
nà-tans: floating, swimming

163

nauseò-sus: nauseous
nuviculà-ris: pertaining to a ship
neapolità-nus: Neapolitan
nebulò-sus: nebulous, clouded,
 obscure
negléc-tus: neglected, overlooked
nelumbifò-lius: nelumbo-leaved
nemorà-lis, nemorò-sus: of
 groves or woods
nepetoì-des: nepeta-like
nephról-epis: kidney-scale
nereifò-lius, neriifo-lius: olean-
 der-leaved
nervò-sus: nerved
níc-titans: blinking, moving
nì-dus: nest
nì-ger: black
nigrà-tus: blackish
nigrés-cens: becoming black
níg-ricans: black
nigricór-nis: black-horned
nigrofrúc-tus: black-fruited
níg-ripes: black-footed
nilót-icus: of the Nile
nippón-icus: of Nippon (Japan)
nì-tens, nít-idus: shining
nivà-lis, nív-eus: snowy, white
nivò-sus: full of snow
nobíl-ior: more noble
nób-ilis: noble, famous
nobilís-simus: very noble
noctiflò-rus: night-flowering
noctúr-nus: of the night
nodiflò-rus: with flowers at
 nodes
nodò-sus: with nodes, jointed
nodulò-sus: with small nodes
nòli-tángere: do not touch,
 touch-me-not
nonscríp-tus: undescribed
norvég-icus: Norwegian
notà-tus: marked
nò-væ-án-gliæ: of New Eng-

land
nò-væ-cæsár-eæ: of New Jersey
nò-væ-zealánd-iæ: of New Zealand
nò-vi-bél-gii: of New York
 (New Belgium)
nubíc-olus: dwelling among
 clouds
nubíg-enus: cloud-born
nucíf-era: nut-bearing
nudà-tus: nude, stripped
nudicaù-lis: naked-stemmed
nudiflò-rus: naked-flowered
nù-dus: nude, naked
numíd-icus: of Numidia
numís-matus: pertaining to
 money
nummularifò-lius: money-leaved
nummulà-rius: money-like
nù-tans: nodding
nyctagín-eus, nyctíc-alus: night-
 blooming
nymphoì-des: nymphea-like

obcón-icus: inversely conical
obcordà-tus: inversely cordate
obè-sus: obese, fat
obfuscà-tus: clouded, confused
oblanceolà-tus: inversely lanceolate
oblì-quus: oblique
obliterà-tus: obliterated, erased
oblongà-tus: oblong
oblongifò-lius: oblong-leaved
oblón-gus: oblong
obovà-tus: inverted ovate, obo-
 vate
obscù-rus: obscure, hidden
obsolè-tus: obsolete, rudimen-
 tary
obtusà-tus: obtuse, blunt
obtusifò-lius: obtuse-leaved

obtusíl-obus: obtuse-lobed
obtù-sior: more obtuse
obtù-sus: obtuse, blunt, rounded
obvallà-tus: apparently walled up
occidentà-lis: western
oceán-icus: oceanic
ocellà-tus: with small eyes
ochnà-ceus: ochna-like
ochrà-ceus: ochre-colored
ochreà-tus: with an ochrea or boot-sheath
ochroleù-cus: yellowish-white
octán-drus: with eight anthers
octopét-alus: eight-petaled
octophýl-lus: eight-leaved
oculà-tus: eyed
ocymoì-des: ocimum like
odessà-nus: of Odessa (southern Russia)
odontì-tes: tooth
odontochì-lus: with toothed lip
odoratís-simus: very fragrant
odorà-tus, odò-rus: odorous, fragrant
officinà-lis: officinal, medicinal
officinà-rum: of the apothecaries
oleæfò-lius, oleifò-lius: olive-leaved
oleíf-era: oil-bearing
oleoì-des: olive-like
olerà-ceus: oleraceous, vegetable-garden herb used in cooking
oligán-thus: few-flowered
oligocár-pus: few-fruited
oligophýl-lus: few-leaved
oligospér-mus: few-seeded
olitò-rius: pertaining to vegetable-gardens
olivà-ceus: olive-like
olivæfór-mis: olive-shaped
olým-picus: of Olympus

omnív-orus: of all kinds of food
onobrychioì-des: onobrychis-like
opà-cus: opaque, shaded
operculà-tus: with a lid
ophiocár-pus: snake-fruit
ophioglossifò-lius: ophioglossum-leaved
ophioglossoì-des: ophioglossum-like
ophiuroì-des: ophiurus-like
oppositiflò-rus: opposite-flowered
oppositifò-lius: opposite-leaved
opuliflò-rus: opulus-flowered
opulifò-lius: opulus-leaved
orbiculà-ris, orbiculà-tus: orbicular, round
orchíd-eus: orchid-like
orchidiflò-rus: orchid-flowered
orchioì-des, orchiò-des: orchid-like
oregà-nus: of Oregon
oreóph-ilus: mountain-loving
orgyà-lis: length of the arms extended, about six feet
orientà-lis: oriental, eastern
origanifò-lius: origanum-leaved
origanoì-des: origanum-like
ór-nans: ornamented or ornamenting
ornatís-simus: very showy
ornà-tus: ornate, adorned
ornithocéph-alus: like a bird's head
ornithóp-odus, ornith-opus: like a bird's foot
ornithorhýn-chus: shaped like a bird's beak
oroboì-des: orobus-like
orthób-otrys: straight-clustered
orthocár-pus: straight-fruited
orthochì-lus: straight-lipped
orthóp-terus: straight-winged

orthosép-alus: straight-sepaled
osmán-thus: fragrant-flowered
ovalifò-lius: oval-leaved
ovà-lis: oval
ovatifò-lius: ovate-leaved
ovà-tus: ovate
ovíf-era, ovíg-era: egg-bearing
ovì-nus: pertaining to sheep
oxyacán-thus: sharp-spined
oxygò-nus: sharp-angled, acute-angled
oxypét-alus: sharp-petaled
oxyphýl-lus: sharp-leaved
oxysép-alus: sharp-sepaled

pabulà-rius: of fodder or pasturage
pachyán-thus: thick-flowered
pachycár-pus: with thick pericarp
pachyneù-rus: thick-nerved
pachyphlœ-us: thick-barked
pachyphýl-lus: thick-leaved
pachýp-terus: thick-winged
pacíf-icus: of the Pacific
palæstì-nus: of Palestine
paleà-ceus: with palea, chaffy
pál-lens: pale
pallés-cens: becoming pale
pallià-tus: cloaked
pallidiflò-rus: pale-flowered
pallidifò-lius: pale-leaved
pallidispì-nus: pale-spined
pál-lidus: pale
palliflà-vens: pale yellow
palmà-ris: palmate
palmatíf-idus: palmately cut
palmà-tus: palmate
palmifò-lius: palm-leaved
paludò-sus, palús-tris: marsh-loving

pandurà-tus: fiddle-shaped
paniculà-tus: paniculate
paniculíg-era: panicle-bearing
pannón-icus: of Pannonia (Hungary)
pannò-sus: ragged, tattered
papaverà-ceus: poppy-like
papillionà-ceus: butterfly-like
papillò-sus: with papillæ or protuberances
papyrà-ceus: papery
papyríf-era: paper-bearing
paradisì-acus: of parks or gardens
paradóx-us: paradoxical, strange
parasít-icus: of a parasite, parasitic
pardalì-nus: leopard-like, spotted
pardì-nus: leopard-spotted
parnassifò-lius: parnassia-leaved
partì-tus: parted
parviflò-rus: small-flowered
parvifò-lius: small-leaved
parvís-simus: very small
pár-vulus: very small
pár-vus: small
patagón-icus: of Patagonia
patavì-nus: of Padua
patellà-ris: circular, disk-shaped
pà-tens: spreading
pát-ulus: spreading
pauciflò-rus: few-flowered
paucifò-lius: few-leaved
paucinér-vis: few-nerved
paupér-culus: poor
pavonì-nus: peacock-like
pectinà-ceus, pectinà-tus: pectinate, comb-like
pectiníf-era: comb-bearing
pectorà-lis: shaped like a breastbone
pedatíf-idus: pedately cut

pedà-tus: footed; bird-footed; palmately divided with side divisions again cleft

pedemontà-nus: of Piedmont (Italy)

pediculà-rius: louse, lousy

pedunculà-ris, pedunculà-tus: peduncled, stalked

pedunculò-sus: with many peduncles

pellù-cidus: with transparent dots

peltà-tus: peltate; shield-shaped

peltifò-lius: peltate-leaved

pelvifór-mis: pelvis-shaped

penduliflò-rus: pendulous-flowered

pendulì-nus: somewhat pendulous

pén-dulus: pendulous, hanging

penicillà-tus: hair-penciled

peninsulà-ris: peninsular

pennà-tus: feathered, pinnate

pennìg era: bearing feathers

penninér-vis: feather-veined

pennsylván-icus: of Pennsylvania

pén-silis: pensile, hanging

pentadè-nius: five-toothed

pentagò-nus: five-angled

pentàg-ynus: of five pistils

pentàn-drus: of five stamens

pentàn-thus: five-flowered

pentàl-ophus: five-winged or five-tufted

pentapetaloì-des: like five petals

pentahýl-lus: five-petaled

pentàp-terus: five-winged

peploì-des: peplis-like

perbél-lus: very beautiful

percús-sus: sharp-pointed

peregrì-nus: exotic, foreign

perén-nans, perén-nis: perennial

perfolià-tus: perfoliate, with leaf surrounding the stem

perforà-tus: perforated, with holes

perfós-sus: perfoliate

pergrác-ilis: very slender

permíx-tus: much mixed

persicæfò-lius, persicifò-lius: peach-leaved

pér-sicus: of Persia; also the peach

persís-tens: persistent

perspíc-uus: clear, transparent

pertù-sus: thrust through, perforated

perulà-tus: pocket-like

peruvià-nus: Peruvian

petaloíd-eus: petal-like

petiolà-ris, petiolà-tus: petioled

petræ-us: rock-loving

petrocál-lis: rock beauty

phæocár-pus: dark-fruited

phæ-us: dusky

philadél phicus: of the Philadelphia region

philoxeroì-des: philoxera-like

phillyræoì-des: phillyrea-like

phleioì-des: phleum-like (Phleum: Gramineæ)

phlogiflò-rus: flame-flowered, phlox-flowered

phlogifò-lius: phlox-leaved

phœníc-eus: purple-red

phœnicolà-sius: purple-haired

phrýg-ius: of Phrygia (Asia Minor)

phyllanthoì-des: phyllanthus-like

phyllomanì-acus: running wildly to leaves

phymatochì-lus: long-lipped

phytolaccoì-des: phytolacca-like

167

picturà-tus: painted-leaved, pictured, variegated

píc-tus: painted

pileà-tus: with a cap

pilíf-era: bearing soft hairs

pilosiús-culus: slightly pilose

pilò-sus: pilose, shaggy, with soft hairs

pilulà-ris: fruit globular

pilulíf-era: globule-bearing

pimeleoì-des: pimella-like

pimpinellifò-lius: pimpinella-leaved

pinetò-rum: of pine forests

pín-eus: of the pine

pinguifò-lius: fat-leaved

pinifò-lius: pine-leaved

pinnatíf-idus: pinnately cut

pinnatifò-lius: pinnate-leaved

pinnát-ifrons: pinnate-fronded

pinnatinér-vis: pinnate-nerved

pinnà-tus: pinnate

piperì-ta: peppermint-scented

pisíf-era: pea-bearing

pisocár-pus: pea-fruited

placà-tus: quiet, calm

placentifór-mis: quoit-shaped

planiflò-rus: flat-flowered

planifò-lius: flat-leaved

plán-ipes: flat-footed

plantagín-eus: plantain-like

plà-nus: plane, flat

platanifò-lius: platanus-leaved

planatoì-des: platanus-like

platán-thus: broad-flowered

platycán-thus: broad-spined

platycár-pus: broad-fruited

platycaù-lon: broad-stemmed

platycén-tra: broad-centered

platýc-ladus: broad-branched

platyglós-sus: broad-tongued

platyneù-rus: broad-nerved

platypét-alus: broad-petaled

platyphýl-lus: broad-leaved

platýp-odus, plát-ypus: broad-footed or -stalked

platýs-pathus: broad-spathed

platyspér-mus: broad-seeded

pleioneù-rus: more- or many-nerved

pleniflò-rus: double-flowered

plenís-simus: very full or double

plè-nus: full, double

pleurós-tachys: side-spiked

plicà-tus: plicate, plaited

plumà-rius, plumà-tus: plumed, feathered

plumbaginoì-des: plumbago-like

plúm-beus: of lead

plumò-sus: feathery

pluriflò-rus: many-flowered

poculifór-mis: deep cup-shaped

podág-ricus: gouty-stalked

podalyriæfò-lius: podalyria-leaved

podocár-pus: with stalked fruits

podól-icus: of Podolia (southwestern Russia)

podophýl-lus: with stalked leaves

poét-icus: pertaining to poets

polifò-lius: polium-leaved, white-leaved

polì-tus: polished

polyacán-thus: many-spined

polyán-drus: with many stamens

polyán-themos, polyán-thus: many-flowered

polybót-rya: many-clustered

polybúl-bon: with many bulbs

polycár-pus: many-fruited

polycéph-alus: many-headed

polychrò-mus: many-colored

polydác-tylus: many-fingered

polygaloì-des: polygala-like

polýg-amus: polygamous, sexes mixed

polýl-epis: with many scales

polýl-ophus: many-crested

polymór-phus: of many forms, variable

polyét-alus: many-petaled

polyphýl-lus: many-leaved

polyrrhì-zus: many-rooted

polysép-alus: many-sepaled

polyspér-mus: many-seeded

polystà-chyus: many-spiked

polystíc-tus: many-dotted

pomà-ceus: pome-like

pomeridià-nus: afternoon

pomíf-era: pome-bearing

pompò-nius: of a tuft or topknot

ponderò-sus: ponderous, heavy

pón-ticus: of Pontus (Asia Minor)

populifò-lius: poplar-leaved

popúl-neus: pertaining to poplars

porcì nus: pertaining to swine

porophýl-lus: porum-leaved, leek-leaved

porphỳ-reus: purple

porphyroneù-rus: purple-nerved

porphyrostè-le: purple-columned

porrifò-lius: porrum- or leek-leaved

portulà-ceus: portulaca-like

potamóph-ilus: swamp-loving, river-loving

potatò-rum: of the drinkers

prœál-tus: very tall

prœ̀-cox: precocious, very early

prœmór-sus: bitten at the end

prœ̀-stans: distinguished, excelling

prœtéx-tus: bordered

prasinà-tus: greenish

prás-inus: grass-green

pratén-sis: of meadows

pravís-simus: very crooked

precatò-rius: praying, prayerful

primulœfò-lius, primulifò-lius: primrose-leaved

primúl-inus: primrose-like

primuloì-des: primrose-like

prín-ceps: princely, first

prismát-icus: prismatic, prism-shaped

prismatocár-pus: prism-fruited

proboscíd-eus: proboscis-like

procè-rus: tall

procúm-bens: procumbent

procúr-rens: extending

prodúc-tus: produced, lengthened

profù-sus: profuse

prolíf-era: producing offshoots

prolíf-icus: prolific, fruitful

propén-dens: hanging down

propín-quus: related, near to

prostrà-tus: prostrate

protrù sus: protruding

provincià-lis: provincial

pruinà-tus, pruinò-sus: with a hoary bloom

prunelloì-des: prunella-like

prunifò-lius: plum-leaved

prù-riens: itching

psilostè-mon: slender- or naked-stamened

psittác-inus: parrot-like

psittacò-rum: of the parrots

psycò-des: fragrant

ptarmicœfò-lius: ptarmica-leaved

ptarmicoì-des: ptarmica-like

pterán-thus: with winged flowers

pteridoì-des: pteris-like

pteroneù-rus: winged-nerved

pù-bens: downy

puberulén-tus, pubér-ulus: somewhat pubescent
pubés-cens: pubescent, downy
pubíg-era: down-bearing
pubiflò-rus: pubescent-flowered
pubinér-vis: pubescent-nerved
pudì-cus: bashful, retiring, shrinking
pugionifór-mis: dagger-formed
pulchél-lus: pretty, beautiful
púl-cher: handsome, beautiful
pulchér-rimus: very handsome
púl-lus: dark colored, dusky
pulverulén-tus: powdered, dust-covered
pulvinà-tus: cushion-like
pù-milus: dwarf
punctatís-simus: very spotted
punctà-tus: punctate, dotted
punctilób-ulus: dotted-lobed
pún-gens: piercing, sharp-pointed
puníc-eus: reddish-purple
púr-gans: purging
purpurà-ceus: purple
purpurás-cens: becoming purple
purpurà-tus, purpù-reus: purple
pusíl-lus: very small
pustulà-tus: as though blistered
pycnacán-thus: densely spined
pycnán-thus: densely flowered
pycnocéph-alus: thick-headed
pycnostà-chyus: thick-spiked
pygmæ̀-us: pigmy
pyramidà-lis: pyramidal
pyrenæ̀-us, pyrenà-icus: of the Pyrenees
pyrifò-lius: pear-leaved
pyrifór-mis: pear-shaped
pyxidà-tus: box-like

quadrangulà-ris, quadrangulà-tus: four-angled
quadrà-tus: in four or fours
quadriaurì-tus: four-eared
quadríc-olor: of four colors
quadridentà-tus: four-toothed
quadríf-idus: four-cut
quadrifò-lius: four-leaved
quadripartì-tus: four-parted
quadrivál-vis: four-valved
quadrivúl-nerus: four-wounded
quercifò-lius: oak-leaved
quérc-inus: of the oak
quinà-tus: in fives
quinquéc-olor: of five colors
quinqueflò-rus: five-flowered
quinquefò-lius: five-leaved
quinqueloculà-ris: five-celled
quinquenér-vis: five-nerved
quinquepunctà-tus: five-spotted
quinquevúl-nerus: five-wounded or -marked

racemiflò-rus: raceme-flowered
racemò-sus: flowers in racemes
rà-dians: radiating
radià-tus: radiate, rayed
radì-cans: rooting
radicà-tus: having roots
radicò-sus: many-rooted
radì-cum: of roots
radiò-sus: with many rays
rád-ula: rough, like a scraper
ramentà-ceus: bearing a hair-like covering
ramiflò-rus: with branching inflorescence
ramondioì-des: ramondia-like
ramosís-simus: much-branched
ramò-sus: branched
ramulò-sus: having many branchlets
raníf-era, frog-bearing

ranunculoì-des: ranunculus-like
rapà-ceus: pertaining to turnips
rapunculoì-des: rapunculus-like
rariflò-rus: scattered-flowered
rà-rus: rare, uncommon
ràu-cus: hoarse, raw
reclinà-tus: reclined, bent back
réc-tus: straight, upright
recurvà-tus: recurved
recurvifò-lius: recurved-leaved
recúr-vus: recurved
redivì-vus: restored, brought to life
reduplicà-tus: duplicated again
refléx-us: reflexed, bent back
refrác-tus: broken
refúl-gens: brightly shining
regà-lis: regal, royal
regér-minans: re-germinating
Regì-na: queen
rè-gius: regal, royal, kingly
religiò-sus: used for religious purposes
remotiflò-rus: distantly flowered
remò-tus: remote, with parts distant
renifór-mis: kidney-shaped
repán-dus: with margin wavy
rè-pens: creeping
replicà-tus: folded back
rép-tans: creeping
reséc-tus: cut off
resiníf-era: resin-bearing
resinò-sus: full of resin
reticulà-tus: reticulate, netted
retinò-des: retained
retór-tus: twisted back
retrofléx-us: reflexed
retrofrác-tus: broken or bent backwards
retù-sus: retuse, notched slightly at a rounded apex
revér-sus: reversed

revolù-tus: revolute, rolled backwards
Réx: king
rhamnifò-lius: rhamnus-leaved
rhamnoì-des: rhamnus-like
rhexifò-lius: rhexia-leaved
rhipsalioì-des: rhipsalis-like
rhizophýl-lus: root-leaved, leaves rooting
rhodán-thus: rose-flowered
rhodochì-lus: rose-lipped
rhodocínc-tus: rose-girdled
rhodoneù-rus: rose-nerved
rhoifò-lius: rhœas-leaved
rhomboíd-eus: rhomboidal
rhóm-beus: rhombic
rhytidophýl-lus: wrinkle-leaved
ricinifò-lius: ricinus-leaved
ricinoì-des: ricinus-like
rì-gens: rigid, stiff
rigidís-simus: very rigid
rigíd-ulus: somewhat rigid
ríg-idus: rigid, stiff
rín-gens: gaping
ripà-rius: of river banks
rivà-lis: pertaining to brooks
rivulà-ris: brook-loving
robustispì-nus: stout-spined
robús-tus: robust, stout
romà-nus: Roman
rosà-ceus: rose-like
rosæflò-rus: rose-flowered
rò-seus: rose, rosy
rosmarinifò-lius: rosemary-leaved
rostrà-tus: rostrate, beaked
rosulà-ris: in rosettes
rotà-tus: wheel-shaped
rotundà-tus: rotund
rotundifò-lius: round-leaved
rotún-dus: rotund, round
rubellì-nus, rubél-lus: reddish
rù-bens, rù-ber: red, ruddy

rubér-rimus: very red
rubés-cens: becoming red
rubicún-dus: rubicund, red
rubiginò-sus: rusty
rubioì-des: rubia-like
rubríc-alyx: calyx red
rubricaù-lis: red-stemmed
rubrifò-lius: red-leaved
rubronér-vis: red-veined
rù-dis: wild, not tilled
rudiús-culus: wild, wildish
rufés-cens: becoming red
rufíd-ulus: somewhat rufid, reddish
rufinér-vis: red-nerved
rù-fus: red, reddish
rugò-sus: rugose, wrinkled
runcinà-tus: runcinate
rupíf-ragus: rock-breaking
rupés-tris: rock-loving
rupíc-olus: growing on cliffs or ledges
ruscifò-lius: ruscus-leaved
russà-tus: reddish, russet
rusticà-nus, rús-ticus: rustic, pertaining to the country
ruthén-icus: Ruthenian (Russian)
rutidobúl-bon: rough-bulbed
rutifò-lius: ruta-leaved
rù-tilans: red, becoming red

saccà-tus: saccate, bag-like
saccharà-tus: containing sugar, sweet
saccharíf-era: sugar-bearing
sacchár-inus: saccharine
saccharoì-des: like sugar
sác-charum: of sugar
saccíf-era: bag-bearing
sacrò-rum: sacred, of sacred places

sagittà-lis, sagittù-tus: sagittate, arrow-like
sagittifò-lius: arrow-leaved
salicariæfò-lius: willow-leaved
salicifò-lius: willow-leaved
salíc-inus: willow-like
salicornioì-des: salicornia-like
salíg-nus: of the willow
salì-nus: salty, of salty places
salsuginò-sus: salt-marsh-loving
salviæfò-lius, salvifò-lius: salvia-leaved
sambucifò-lius: sambucus-leaved, elder-leaved
sambucì-nus: sambucus- or elder-like
sánc-tus: holy
sanguín-eus: bloody, blood-red
sáp-idus: savory, pleasing to taste
sapién-tum: of the wise men or authors
saponà-ceus: soapy
sarcò-des: flesh-like
sarmát-icus: of Sarmatia; Russian
sarmentò-sus: bearing runners
satì-vus: cultivated
saturà-tus: saturated
saurocéph-alus: lizard-headed
saxát-ilis: found among rocks
saxíc-olus: growing among rocks
saxò-sus: full of rocks
scà-ber: scabrous, rough
scabér-rimus: very rough
scabiosæfò-lius: scabiosa-leaved
scabrél-lus, scáb-ridus: somewhat rough
scán-dens: scandent, climbing
scapò-sus: with scapes
scariò-sus: scarious, thin and not green
scép-trum: of a scepter
schidíg-era: spine-bearing

schistò-sus: schistose
schizoneù-rus: cut-nerved
schizopét-alus: cut-petaled
schizophýl-lus: cut-leaved
scholà-ris: pertaining to a school
scilloì-des: squill-like
sclerocár-pus: hard-fruited
sclerophýl-lus: hard-leaved
scopà-rius: broom or broom-like
scopulò-rum: of the rocks
scorpioì-des: scorpion-like
scorzoneroì-des: scorzonera-like
scól-ica: Scotch
scúl-ptus: carved
scutellà-ris, scutellà-tus: salver- or dish-shaped
scutà-tus: buckler-shaped
scù-tum: a shield
sebíf-era: tallow-bearing
sebò-sus: full of tallow or grease
sechellà-rum: of the Seychelles (Indian Ocean)
seclù-sus: hidden, secluded
secundiflò-rus: secund-flowered
secún-dus, secundà-tus: secund, side-flowering
securíg-era: axe-bearing
ség-etum: of cornfields
selaginoì-des: selago-like, club-moss-like
semialà-tus: semi-winged
semibaccà-tus: semi-berried
semicaudà-tus: semi-tailed
semicylín-dricus: semi-cylindrical
semidecán-drus: half ten-stamened
semipinnà-tus: imperfectly pinnate
semperflò-rens: ever flowering
sempér-virens: ever green
sempervivoì-des: sempervivum-like

senecioì-des: senecio-like
senì-lis: senile, old, white-haired
sensíb-ilis: sensitive
sensití-vus: sensitive
sepià-rius: of or pertaining to hedges
sè-pium: of hedges or fences
septangulà-ris: seven-angled
septém-fidus: seven-cut
septém-lobus: seven-lobed
septempunctà-tus: seven-spotted
septentrionà-lis: northern
sepúl-tus: sepulchered, interred
sericán-thus: silky-flowered
seríc-eus: silky
sericíf-era, sericóf-era: silk-bearing
serót-inus: late, late-flowering or late-ripening
sér-pens: creeping, crawling
serpentì-nus: of snakes, serpentine
serpyllifò-lius: thyme-leaved, serpyllum-leaved
serratifò-lius: serrate-leaved
serrà-tus: serrate, saw-toothed
serrulà-tus: somewhat serrate
sesquipedà-lis: one foot and a half long or high
sessiflò-rus: sessile-flowered
sessifò-lius: sessile-leaved
sessiliflò-rus: sessile-flowered
sessilifò-lius: sessile-leaved
sés-silis: sessile, stalkless
setà-ceus: bristle-like
setifò-lius: bristle-leaved
setíg-era, sét-iger: bristle-bearing
setíp-odus: bristle-footed
setispì-nus: bristle-spined
setò-sus: full of bristles
setulò-sus: full of small bristles
sexangulà-ris: six-angled

sià-meus: of Siam
sibír-icus: of Siberia
siculifór-mis: dagger-formed
síc-ulus: of Sicily
siderophloì-us: iron bark
sideróx-ylon: iron wood
signà-tus: marked, designated
silaifò-lius: silaus-leaved
silíc-eus: pertaining to or growing in sand
siliculò-sus: bearing silicles
siliquò-sus: bearing siliques
silvát-icus, silvés-tris: pertaining to woods
sím-ilis: similar, like
sím-plex: simple, unbranched
simplicicaù-lis: simple-stemmed
simplicifò-lius: simple-leaved
simplicís-simus: simplest
sím-ulans: similar to, resembling
sín-icus: Chinese
sinuà-tus, sinuò-sus: sinuate, wavy-margined
siphilít-icus: syphilitic
sisalà-nus: pertaining to sisal
sisymbrifò-lius: sisymbrium-leaved
smarág-dinus: of emerald
smilác-inus: of smilax
sobolíf-era: bearing creeping rooting stems or roots
socià-lis: sociable, companionable
socotrà-nus: of Socotra (island off Arabia)
sodomè-um: of Sodom
solandriflò-rus: solandra-flowered
solà-ris: of the sun
soldanelloì-des: like soldanella
sól-idus: solid, dense
somníf-era: sleep-producing
sonchifò-lius: sonchus-leaved

sorbifò-lius: sorbus-leaved
sór-didus: dirty
spadíc-eus: with a spadix
sparsiflò-rus: sparsely flowered
sparsifò-lius: sparsely leaved
spár-sus: sparse, few
spárteus: pertaining to the broom or Spartium
spathà-ceus: with a spathe
spathulà-tus: spatulate, spoon-shaped
spathulifò-lius: spatulate-leaved
speciosís-simus: very showy
speciò-sus: showy, good-looking
spectáb-ilis: spectacular, remarkable, showy
spectán-drus: showy
spéc-trum: an image, apparition
speculà-tus: shining, as if with mirrors
sphacelà-tus: dead, withered, diseased
sphǽr-icus: spherical
sphǣcocár-pus: spherical-fruited
sphǣrocéph-alus: spherical-headed
sphǣroíd-eus: sphere-like
sphǣrostà-chyus: spherical-spiked
spicà-tus: spicate, with spikes
spicifór-mis: spike-shaped
spicíg-era: spike-bearing
spiculifò-lius: spicule-leaved
spinà-rum: spiny
spinés-cens: somewhat spiny
spiníf-era: bearing spines
spinosís-simus: very spiny
spinò-sus: full of spines
spinulíf-era: bearing small spines
spinulò-sus: somewhat or weakly spiny
spirà-lis: spiral
spirél-lus: little spiral

splén-dens: splendid
splendidís-simus: very splendid
splén-didus: splendid
spondioì-des: spondias-like
   (Spondias: Anacardiaceæ)
spumà-rius: frothing
spù-rius: spurious, false
squà-lens, squál-idus: squalid, filthy
squamà-tus: squamate, with small scale-like leaves or bracts
squamò-sus: full of scales
squarrò-sus: with parts spreading or even recurved at ends
stachyoì-des: stachys-like
stamín-eus: bearing prominent stamens
stáns: standing, erect, upright
stauracán-thus: with spines cross-shaped
stellà-ris, stellà-tus: stellate, starry
stellíp-ilus: with stellate hairs
stellulà-tus: somewhat stellate
stenocár-pus: narrow-fruited
stenocéph-alus: narrow-headed
stenóg-ynus: with narrow stigma
stenopét-alus: narrow-petaled
stenophýl-lus: narrow-leaved
stenóp-terus: narrow-winged
stenostà-chyus: narrow-spiked
stér-ilis: sterile, infertile
stigmát-icus: marked, of stigmas
stigmò-sus: much marked, pertaining to stigmas
stipulà-ceus, stipulà-ris, stipulà-tus: having stipules
stipulò-sus: having large stipules
stoloníf-era: bearing stolons or runners that take root
stramineofrúc-tus: with straw-colored fruit

stramín-eus: straw-colored
strangulà-tus: strangled, constricted
streptocár-pus: twisted-fruited
streptopét-alus: petals twisted
streptophýl-lus: twisted-leaved
streptosép-alus: sepals twisted
striát-ulus: faintly striped
strià-tus: striated, striped
strictiflò-rus: stiff-flowered
stríc-tus: strict, upright, erect
strigillò-sus: somewhat strigose
strigò-sus: strigose
strigulò-sus: with small or weak appressed hairs
striolà-tus: faintly striped
strobilà-ceus: resembling a cone
strobilíf-era: cone-bearing
strumà-rius: of tumors or ulcers
strumà-tus: with tumors or ulcers
strumò-sus: having cushion-like swellings
stylò-sus: with prominent styles
styphelioì-des: styphelia-like
styracíf-luus: flowing with storax or gum
suavè-olens: sweet-scented
suà-vis: sweet, agreeable
suavís-simus: sweetest
subacaù-lis: somewhat stemmed
subalpì-nus: nearly alpine
subauriculà-tus: somewhat eared
subcærù-leus: slightly blue
subcà-nus: somewhat hoary
subcarnò-sus: rather fleshy
subcordà-tus: somewhat cordate
subdivaricà-tus: slightly divaricate
subedentà-tus: nearly toothless
suberculà-tus: of cork, corky
suberéc-tus: somewhat erect
suberò-sus: cork-barked

subfalcà-tus: somewhat falcate
subglaù-cus: somewhat glaucous
subhirtél-lus: somewhat hairy
sublunà-tus: somewhat crescent-shaped
submér-sus: submerged
subperén-nis: nearly perennial
subpetiolà-tus: partially petioled
subscán-dens: partially climbing
subsés-silis: nearly sessile
subsinuà-tus: somewhat sinuate
subterrà-neus: underground
subulà-tus: awl-shaped
subumbellà-tus: somewhat umbellate
subvillò-sus: somewhat soft-hairy
subvolù-bilis: somewhat twining
succotrì-nus: of Socotra; see socotranus
succulén-tus: succulent, fleshy
suéc-icus: Swedish
suffrutés-cens, suffruticò-sus: somewhat shrubby
suffúl-tus: supported
sulcà-tus: sulcate, furrowed
sulphù-reus: sulfur-colored
sumatrà-nus: of Sumatra
supér-biens, supér-bus: superb, proud
supercilià-ris: eyebrow-like
supér-fluus: superfluous, redundant
supì-nus: prostrate
supraaxillà-ris: above the axils
supracà-nus: gray-pubescent above
surculò-sus: producing suckers
susià-nus: of Susa, an ancient city of Persia
suspén-sus: suspended, hung
sylvát-icus: forest-loving
sylvés-ter, sylvés-tris: of woods or forests
sylvíc-olus: growing in woods
syphilít-icus: syphilitic
syrì-acus: Syrian
syringán-thus: syringa-flowered
syringifò-lius: syringa-leaved

tabulæfór-mis, tabulifór-mis: table-formed
tabulà-ris: table-like, flattened horizontally
tædíg-era: cone-bearing, torch-bearing
tanacetifò-lius: tansy-leaved
taraxicifò-lius: dandelion-leaved
tardiflò-rus: late-flowered
tardì-vus: tardy, late
tartà-reus: with a loose or rough crumbling surface
tatár-icus: of Tartary
taù-reus: of oxen
taù-ricus: Taurian, Crimean
taurì-nus: bull-like, ox-like, pertaining to cattle
taxifò-lius: yew-leaved
téch-nicus: technical, special
tectò-rum: of roofs or houses
téc-tus: concealed, covered
tellimoì-des: tellima-like (Tellima: Saxifrageæ)
temulén-tus: drunken
tenacís-simus: most tenacious
tè-nax: tenacious, strong
tenebrò-sus: of dark or shaded places
tenél-lus: slender, tender, soft
tè-ner, tén-era: slender, tender, soft
tentaculà-tus: with tentacles
tenuicaù-lis: slender-stemmed
tenuiflò-rus: slender-flowered
tenuifò-lius: slender-leaved
tenuíl-obus: slender-lobed

tenù-ior: more slender
tenuipét-alus: slender-petaled
tén-uis: slender, thin
tenuís-simus: very slender
tenuistỳ-lus: slender-styled
terebinthà-ceus: of turpentine
terebinthifò-lius: terebinthus-leaved
terebínth-inus: of turpentine
tè-res: terete, circular in cross-section
teretifò-lius: terete-leaved
tereticór-nis: with terete horns
terminà-lis: terminal
ternatè-a: of island of Ternate in Moluccas
ternà-tus: in threes
ternifò-lius: leaves in threes
terrés-tris: of the earth
tessellà-tus: tessellate, checkered
testà-ceus: light brown, brick-colored; also testaceous
testiculà-tus: testiculated, testicled
testudinà-rius: like a tortoise-shell
tetracán-thus: four-spined
tetragonól-obus: with four-angled pod
tetragò-nus: four-angled
tetrám-erus: of four members
tetrán-drus: four-anthered
tetrán-thus: four-flowered
tetraphýl-lus: four-leaved
tetráp-terus: four-winged
tetraquè-trus: four-cornered
teucrioì-des: teucrium-like
texà-nus: of Texas, Texan
téx-tilis: textile, woven
thapsoì-des: thapsus-like, mullein-like
thalictroì-des: thalictrum-like
thebà-icus: of Thebes

theíf-era: tea-bearing
thermà-lis: warm, of warm springs
thibét-icus: of Tibet
thuríf-era: incense-bearing
thuyoì-des, thyoì-des: thuja-like
thymifò-lius: thyme-leaved
thymoì-des: thyme-like
thyrsiflò-rus: thyrse-flowered
thrysoì-des: thyrse-like
tibét-icus: of Tibet
tibíc-inis: of a flute player
tigrì-nus: tiger-striped
tilià-ceus: tilia-like, linden-like
tiliæfò-lius: tilia-leaved
tinctò-rius: belonging to dyers, of dyes
tínc-tus: dyed
tingità-nus: of Tangiers
tipulifór-mis: of the shape of a daddy-long-legs
tità-nus: very large
tomentò-sus: tomentose, densely woolly
tón-sus: clipt, sheared
torminà-lis: useful against colic
torò-sus: cylindrical with contractions at intervals
tortifò-lius: leaves twisted
tór-tilis: twisted
tortuò-sus: much twisted
tór-tus: twisted
torulò-sus: somewhat torose; see *torosus*
toxicà-rius, tóx-icus: poisonous
toxíf-era: poison-producing
trachypleù-ra: rough-ribbed or -nerved
trachyspér-mus: rough-seeded
tragophýl-lus: tragus-leaved
translù-cens: translucent
transpà-rens: transparent
transylván-icus: of Transylvania

177

trapezifór-mis: with four unequal sides
trapezioì-des: trapezium-like
tremuloì-des: like tremulus, the
trembling poplar
trém-ulus: quivering, trembling
triacanthóph-orus: bearing three
spines
triacán-thus: three-spined
trián-drus: with three anthers or
stamens
triangulà-ris, triangulà-tus:
three-angled
trián-gulus: three-angular
tricaudà-tus: three-tailed
tricéph-alus: three-headed
trichóc-alyx: calyx hairy
trichocár-pus: hairy-fruited
trichomanefò-lius: trichomanes-
leaved
trichomanoì-des: trichomanes-
like
trichophýl-lus: hairy-leaved
trichosán-thus: hairy-flowered
trichospér-mus: hairy-seeded
trichót-omus: three-branched or
-forked
tricóc-cus: three-seeded, three-
berried
tríc-olor: three-colored
tricór-nis: three-horned
tricuspidà-tus: having three
points
tridác-tylus: three-fingered
trì-dens, tridentà-tus: three-
toothed
trifascià-tus: three-banded
tríf-idus: three-parted
triflò-rus: three-flowered
trifolià-tus: three-leaved
trifoliolà-tus: of three leaflets
trifò-lius: three-leaved

trifurcà-tus, trifúr-cus: three-
forked
triglochidià-tus: with three
barbed bristles
trigonophýl-lus: three-cornered-
leaved; trigonus-leaved
trilineà-tus: three-lined
trilobà-tus, tríl-obus: three-
lobed
trimés-tris: of three months
trinér-vis: three-nerved
trinotà-tus: three-marked or
-spotted
triornithóph-orus: bearing three
birds
tripartì-tus: three-parted
tripét-alus: three-petaled
triphýl-lus: three-leaved
tríp-terus: three-winged
tripunctà-tus: three-spotted
triquè-tris: three-cornered
trispér-mus: three-seeded
tristà-chyus: three-spiked
trís-tis: sad, bitter, dull
triternà-tus: thrice in threes
triúm-phans: triumphant
trivià-lis: common, ordinary
trolliifò-lius: trollius-leaved
tróp-icus: of the tropics
truncát-ulus: somewhat truncate
truncà-tus: truncate, cut off
square
tubæfór-mis: trumpet-shaped
tubà-tus: trumpet-shaped
tuberculà-tus, tuberculò-sus:
having tubercles
tuberò-sus: tuberous
tubíf-era: tube-bearing
tubiflò-rus: trumpet-flowered
tubís-pathus: tube-spathed
tubulò-sus: with tubes
tulipíf-era: tulip-bearing
tù-midus: swollen

turbinà-tus: top-shaped
turbinél-lus: little top-shaped
túr-gidus: turgid, inflated, full
typhì-nus: pertaining to fever
týp-icus: typical

ulíc-inus: like ulex
uliginò-sus: of wet or marshy places
ulmifò-lius: elm-leaved, ulmus-leaved
ulmoì-des: elm-like
umbellà-tus: with umbels
umbellulà-tus: with umbellets
umbonà-tus: bearing at center an umbo or stout projection
umbraculíf-era: umbrella-bearing
umbrò-sus: shaded, shade-loving
uncinà-tus: hooked at the point
undà-tus: waved
undulatifò-lius: undulate-leaved
undulà-tus: undulated, wavy
undulifò-lius: wavy-leaved
unguiculà-ris, unguiculà-tus: clawed
unguipét-alus: petals clawed
unguispì-nus: claw-spined
uníc-olor: one-colored
unicór-nis: one-horned
unidentà-tus: one-toothed
uniflò-rus: one-flowered
unifò-lius: one-leaved
unilaterà-lis: one-sided
unioloì-des: uniola-like
univittà-tus: one-striped
urbà-nus: city-loving
urceolà-tus: urn-shaped
ù-rens: burning, stinging
urentís-simus: very burning or stinging
urníg-era: pitcher-bearing

urophýl-lus: tail-leaved
urostà-chyus: tail-spiked
ursì-nus: pertaining to bears, northern (under the Great Bear)
urticæfò-lius, urticifò-lius: nettle-leaved
urticoì-des: nettle-like
usitatís-simus: most useful
usneoì-des: usnea-like
ustulà-tus: burnt, sere
ù-tilis: useful
utilís-simus: most useful
utriculà-tus: with a small bladdery one-seeded fruit
utriculò-sus: utricled
uvíf-era: grape-bearing

vaccinifò-lius: vaccinium-leaved
vaccinoì-des: vaccinium-like
vacíl-lans: swaying
và-gans: wandering
vaginà-lis. vaginà-tus: sheathed
valdivià-nus: of Valdivia (Chile)
valentì-nus: of Valentia (Spain)
vál-idus: strong
vandà-rum: of vanda (an orchid)
variáb-ilis, và-rians, varià-tus: variable
varicò-sus: varicose
variegà-tus: variegated
variifò-lius: variable-leaved
variifór-mis: of variable forms
và-rius: various, diverse
vegetà-tus, vég-etus: vigorous
velà-ris: pertaining to curtains or veils
velù-tinus: velvety
vè-lox: rapidly growing, swift
venenà-tus: poisonous
venò-sus: veiny

ventricò-sus: ventricose
venús-tus: handsome, charming
verbascifò-lius: verbascum-leaved
verecún-dus: modest, blushing
vermiculà-tus: worm-like
vernà-lis: of spring
vernicíf-era, vernicíf-lua: varnish-bearing
vernicò-sus: varnished
vér-nus: of spring
verrucò-sus: verrucose, warted
verruculò-sus: very warty
versíc-olor: variously colored
verticillà-ris, verticillà-tus: verticillate, whorled
vè-rus: the true or genuine or standard
vés-cus: weak, thin, feeble
vesciculò-sus: with little bladders
vespertì-nus: of the evening, western
vestì-tus: covered, clothed
véx-ans: puzzling, vexatious
vexillà-rius: of the standard petal
viburnifò-lius: viburnum-leaved
viciæfò-lius, vicifò-lius: vetch-leaved
victorià-lis: of Victoria
villò-sus: villous, soft-hairy
viminà-lis, vimín-eus: of osiers
viníf-era: wine-bearing
vinò-sus: full of wine
violà-ceus: violet
violés-cens: becoming violet-colored
vì-rens: green
virés-cens: becoming green
virgà-tus: twiggy
virginà-lis, virgín-eus: virgin
virginià-nus, virgín-icus,

virginién-sis: of Virginia
viridés-cens: becoming green
viridicarinà-tus: green-keeled
viridiflò-rus: green-flowered
viridifò-lius: green-leaved
viridifús-cus: green-brown
vír-idis: green
viridís-simus: very green
viríd-ulus: greenish
viscíd-ulus: somewhat sticky
vís-cidus: viscid, sticky
viscosís-simus: very sticky
viscò-sus: sticky
vità-ceus: vitis-like, vine-like
vitellì-nus: dull yellow approaching red
viticulò-sus: sarmentose
vitifò-lius: grape-leaved
vittà-tus: striped
vittíg-era: bearing stripes
vivíp-arus: producing the young alive, freely producing asexual propagating parts
volgár-icus: of the Volga River
volù-bilis: twining
volù-tus: rolled-leaved
vomitò-rius: emetic
vulcán-icus: of Vulcan or a volcano
vulgà-ris, vulgà-tus: vulgar, common
vulpì-nus: of the fox

wolgár-icus: of the Volga River region; see *volgaricus*

xanthacán-thus: yellow-spined
xánth-inus: yellow
xanthocár-pus: yellow-fruited
xantholeù-cus: yellow-white
xanthoneù-rus: yellow-nerved
xanthophýl-lus: yellow-leaved

xanthorrhì-zus: yellow-rooted
xanthóx-ylon: yellow-wooded
xylonacán-thus: woody-spined

zebrì-nus: zebra-striped

zeylán-icus: of Ceylon
zibethì-nus: like the civet-cat, malodorous
zizanioì-des: zizania-like
zonà-lis, zonà-tus: zoned, banded

A CATALOG OF SELECTED
DOVER BOOKS
IN ALL FIELDS OF INTEREST

A CATALOG OF SELECTED DOVER
BOOKS IN ALL FIELDS OF INTEREST

CONCERNING THE SPIRITUAL IN ART, Wassily Kandinsky. Pioneering work by father of abstract art. Thoughts on color theory, nature of art. Analysis of earlier masters. 12 illustrations. 80pp. of text. 5⅜ x 8½.   0-486-23411-8

CELTIC ART: The Methods of Construction, George Bain. Simple geometric techniques for making Celtic interlacements, spirals, Kells-type initials, animals, humans, etc. Over 500 illustrations. 160pp. 9 x 12. (Available in U.S. only.)   0-486-22923-8

AN ATLAS OF ANATOMY FOR ARTISTS, Fritz Schider. Most thorough reference work on art anatomy in the world. Hundreds of illustrations, including selections from works by Vesalius, Leonardo, Goya, Ingres, Michelangelo, others. 593 illustrations. 192pp. 7⅛ x 10¼.   0-486-20241-0

CELTIC HAND STROKE-BY-STROKE (Irish Half-Uncial from "The Book of Kells"): An Arthur Baker Calligraphy Manual, Arthur Baker. Complete guide to creating each letter of the alphabet in distinctive Celtic manner. Covers hand position, strokes, pens, inks, paper, more. Illustrated. 48pp. 8¼ x 11.   0-486-24336-2

EASY ORIGAMI, John Montroll. Charming collection of 32 projects (hat, cup, pelican, piano, swan, many more) specially designed for the novice origami hobbyist. Clearly illustrated easy-to-follow instructions insure that even beginning papercrafters will achieve successful results. 48pp. 8¼ x 11.   0-486-27298-2

BLOOMINGDALE'S ILLUSTRATED 1886 CATALOG: Fashions, Dry Goods and Housewares, Bloomingdale Brothers. Famed merchants' extremely rare catalog depicting about 1,700 products: clothing, housewares, firearms, dry goods, jewelry, more. Invaluable for dating, identifying vintage items. Also, copyright-free graphics for artists, designers. Co-published with Henry Ford Museum & Greenfield Village. 160pp. 8¼ x 11.   0-486-25780-0

THE ART OF WORLDLY WISDOM, Baltasar Gracian. "Think with the few and speak with the many," "Friends are a second existence," and "Be able to forget" are among this 1637 volume's 300 pithy maxims. A perfect source of mental and spiritual refreshment, it can be opened at random and appreciated either in brief or at length. 128pp. 5⅜ x 8½.   0-486-44034-6

JOHNSON'S DICTIONARY: A Modern Selection, Samuel Johnson (E. L. McAdam and George Milne, eds.). This modern version reduces the original 1755 edition's 2,300 pages of definitions and literary examples to a more manageable length, retaining the verbal pleasure and historical curiosity of the original. 480pp. 5³⁄₁₆ x 8¼.   0-486-44089-3

ADVENTURES OF HUCKLEBERRY FINN, Mark Twain, Illustrated by E. W. Kemble. A work of eternal richness and complexity, a source of ongoing critical debate, and a literary landmark, Twain's 1885 masterpiece about a barefoot boy's journey of self-discovery has enthralled readers around the world. This handsome clothbound reproduction of the first edition features all 174 of the original black-and-white illustrations. 368pp. 5⅜ x 8½.   0-486-44322-1

STICKLEY CRAFTSMAN FURNITURE CATALOGS, Gustav Stickley and L. & J. G. Stickley. Beautiful, functional furniture in two authentic catalogs from 1910. 594 illustrations, including 277 photos, show settles, rockers, armchairs, reclining chairs, bookcases, desks, tables. 183pp. 6½ x 9¼. 0-486-23838-5

AMERICAN LOCOMOTIVES IN HISTORIC PHOTOGRAPHS: 1858 to 1949, Ron Ziel (ed.). A rare collection of 126 meticulously detailed official photographs, called "builder portraits," of American locomotives that majestically chronicle the rise of steam locomotive power in America. Introduction. Detailed captions. xi+ 129pp. 9 x 12. 0-486-27393-8

AMERICA'S LIGHTHOUSES: An Illustrated History, Francis Ross Holland, Jr. Delightfully written, profusely illustrated fact-filled survey of over 200 American lighthouses since 1716. History, anecdotes, technological advances, more. 240pp. 8 x 10¾. 0-486-25576-X

TOWARDS A NEW ARCHITECTURE, Le Corbusier. Pioneering manifesto by founder of "International School." Technical and aesthetic theories, views of industry, economics, relation of form to function, "mass-production split" and much more. Profusely illustrated. 320pp. 6⅛ x 9¼. (Available in U.S. only.) 0-486-25023-7

HOW THE OTHER HALF LIVES, Jacob Riis. Famous journalistic record, exposing poverty and degradation of New York slums around 1900, by major social reformer. 100 striking and influential photographs. 233pp. 10 x 7⅞. 0-486-22012-5

FRUIT KEY AND TWIG KEY TO TREES AND SHRUBS, William M. Harlow. One of the handiest and most widely used identification aids. Fruit key covers 120 deciduous and evergreen species; twig key 160 deciduous species. Easily used. Over 300 photographs. 126pp. 5⅜ x 8½. 0-486-20511-8

COMMON BIRD SONGS, Dr. Donald J. Borror. Songs of 60 most common U.S. birds: robins, sparrows, cardinals, bluejays, finches, more–arranged in order of increasing complexity. Up to 9 variations of songs of each species.
Cassette and manual 0-486-99911-4

ORCHIDS AS HOUSE PLANTS, Rebecca Tyson Northen. Grow cattleyas and many other kinds of orchids–in a window, in a case, or under artificial light. 63 illustrations. 148pp. 5⅜ x 8½. 0-486-23261-1

MONSTER MAZES, Dave Phillips. Masterful mazes at four levels of difficulty. Avoid deadly perils and evil creatures to find magical treasures. Solutions for all 32 exciting illustrated puzzles. 48pp. 8¼ x 11. 0-486-26005-4

MOZART'S DON GIOVANNI (DOVER OPERA LIBRETTO SERIES), Wolfgang Amadeus Mozart. Introduced and translated by Ellen H. Bleiler. Standard Italian libretto, with complete English translation. Convenient and thoroughly portable–an ideal companion for reading along with a recording or the performance itself. Introduction. List of characters. Plot summary. 121pp. 5¼ x 8½. 0-486-24944-1

FRANK LLOYD WRIGHT'S DANA HOUSE, Donald Hoffmann. Pictorial essay of residential masterpiece with over 160 interior and exterior photos, plans, elevations, sketches and studies. 128pp. 9¼ x 10¾. 0-486-29120-0

LIGHT AND SHADE: A Classic Approach to Three-Dimensional Drawing, Mrs. Mary P. Merrifield. Handy reference clearly demonstrates principles of light and shade by revealing effects of common daylight, sunshine, and candle or artificial light on geometrical solids. 13 plates. 64pp. 5⅜ x 8½. 0-486-44143-1

ASTROLOGY AND ASTRONOMY: A Pictorial Archive of Signs and Symbols, Ernst and Johanna Lehner. Treasure trove of stories, lore, and myth, accompanied by more than 300 rare illustrations of planets, the Milky Way, signs of the zodiac, comets, meteors, and other astronomical phenomena. 192pp. 8⅜ x 11. 0-486-43981-X

JEWELRY MAKING: Techniques for Metal, Tim McCreight. Easy-to-follow instructions and carefully executed illustrations describe tools and techniques, use of gems and enamels, wire inlay, casting, and other topics. 72 line illustrations and diagrams. 176pp. 8¼ x 10⅞. 0-486-44043-5

MAKING BIRDHOUSES: Easy and Advanced Projects, Gladstone Califf. Easy-to-follow instructions include diagrams for everything from a one-room house for bluebirds to a forty-two-room structure for purple martins. 56 plates; 4 figures. 80pp. 8¾ x 6⅜. 0-486-44183-0

LITTLE BOOK OF LOG CABINS: How to Build and Furnish Them, William S. Wicks. Handy how-to manual, with instructions and illustrations for building cabins in the Adirondack style, fireplaces, stairways, furniture, beamed ceilings, and more. 102 line drawings. 96pp. 8¾ x 6⅜. 0-486-44259-4

THE SEASONS OF AMERICA PAST, Eric Sloane. From "sugaring time" and strawberry picking to Indian summer and fall harvest, a whole year's activities described in charming prose and enhanced with 79 of the author's own illustrations. 160pp. 8¼ x 11. 0-486-44220-9

THE METROPOLIS OF TOMORROW, Hugh Ferriss. Generous, prophetic vision of the metropolis of the future, as perceived in 1929. Powerful illustrations of towering structures, wide avenues, and rooftop parks–all features in many of today's modern cities. 59 illustrations. 144pp. 8¼ x 11. 0-486-43727-2

THE PATH TO ROME, Hilaire Belloc. This 1902 memoir abounds in lively vignettes from a vanished time, recounting a pilgrimage on foot across the Alps and Apennines in order to "see all Europe which the Christian Faith has saved." 77 of the author's original line drawings complement his sparkling prose. 272pp. 5⅜ x 8½. 0-486-44001-X

THE HISTORY OF RASSELAS: Prince of Abissinia, Samuel Johnson. Distinguished English writer attacks eighteenth-century optimism and man's unrealistic estimates of what life has to offer. 112pp. 5⅜ x 8½. 0-486-44094-X

A VOYAGE TO ARCTURUS, David Lindsay. A brilliant flight of pure fancy, where wild creatures crowd the fantastic landscape and demented torturers dominate victims with their bizarre mental powers. 272pp. 5⅜ x 8½. 0-486-44198-9

Paperbound unless otherwise indicated. Available at your book dealer, online at **www.doverpublications.com**, or by writing to Dept. GI, Dover Publications, Inc., 31 East 2nd Street, Mineola, NY 11501. For current price information or for free catalogs (please indicate field of interest), write to Dover Publications or log on to **www.doverpublications.com** and see every Dover book in print. Dover publishes more than 500 books each year on science, elementary and advanced mathematics, biology, music, art, literary history, social sciences, and other areas.